Patrick Moore's Practical

For further volumes:
http://www.springer.com/series/3192

One-Shot Color Astronomical Imaging

In Less Time, for Less Money!

L.A. Kennedy

 Springer

L.A. Kennedy
Clawson, MI
USA

ISSN 1431-9756
ISBN 978-1-4614-3246-3   ISBN 978-1-4614-3247-0 (eBook)
DOI 10.1007/978-1-4614-3247-0
Springer New York Heidelberg Dordrecht London

Library of Congress Control Number: 2012932205

Springer is part of Springer Science+Business Media (www.springer.com)

Preface

Remember when you got your first telescope? You could spend countless hours looking at the Moon, Mars, Jupiter, and everyone's all-time favorite nighttime attraction – the Ringed Planet – Saturn. If you're like most amateur astronomers, the next thing you did was to buy a bigger telescope so you could see even more marvels in the night sky.

As your telescopes got bigger and bigger, you got to see better and clearer views of the Moon, Mars, Jupiter, and Saturn. You may have seen some star formations, open clusters, or the Seven Sister stars that make up the Pleiades. You also probably got to see some smudges that were globular clusters or maybe even a very faint galaxy or two. After that, you were probably left wanting more and wondering what you had to do to be able to see the many night sky objects that you had heard about or had seen pictures of from the Hubble Space Telescope or perhaps in the astronomy magazines that you had read.

If you're like most people you live in a city or in a major suburban area. With all of the city lights, streetlamps, and your local strip malls shining light into the sky all evening, most of the stars and other astronomical wonders are hidden from your telescope's view. Up until recently, only the largest telescope users, at dedicated observatories on mountaintop locations, had the opportunity to see the many wonders of the night sky. Try as you may, even in a dark sky location the number of objects you can see by looking through an amateur-sized telescope can be somewhat limited. This is the point where most people lose their interest in pursuing astronomy as a hobby. After seeing all the objects that they can see using only their relatively small telescopes, the telescope comes out less and less frequently until eventually it gets tucked away in a garage or a basement. There it stays until the telescope and equipment is finally given to a new astronomy aficionado or worse yet, the once-prized telescope gets unloaded at fire-sale prices in a garage sale.

If you're reading this book, however, there is still hope for that old telescope. You are one of the lucky few who have chosen to push your telescope beyond the normal view through the eyepiece. With the combination of digital imaging and your telescope, you will be able to "see" much more of the night sky than you ever imagined. From clear views of colorful nebulae to stunning images of galaxies thousands of light years away, prepare yourself to be amazed at the endless wonders of the night sky that will soon be within your telescope's reach!

Almost as exciting, one of the greatest benefits of digital imaging is that you can now share your passion with other people – without asking them to spend hours outside in the middle of the night for a few quick glances through the telescope eyepiece. Although it's fun to share telescope time with someone, astronomy is often referred to as the loneliest hobby, simply because it takes some real dedication and fortitude to spend the time needed to see the sights in the middle of the night. Now you can take images at night, see them instantly on your computer, and then save them for viewing later, or showing them off at your leisure, any time of the day or night. You can also process the initial images you take in order to bring out the most interesting features. You can enhance the color or the contrast and you can also label your images so those who aren't into astronomy as much as you are will know exactly what they are looking at. Once you have your images processed into a final version you are happy with, you can publish these images on websites, put them on digital photo frames, or display them on any other digital device.

So if your interest is piqued, turn the page and start reading about how you too can take awesome digital images, in less time, with less money…

Clawson, MI, USA   L.A. Kennedy

About the Author

Having discovered an unknown passion for astronomy in college, L. A. Kennedy has been an avid amateur astronomer and astro-imager ever since. In addition to writing this book, he is very active in sharing his knowledge and passion for astrophotography with interested people at schools, colleges, and in private settings. He currently lives and images under the light-polluted skies of southeastern Michigan.

Kennedy is a purchasing and supply-chain management professional by trade, with years of experience in several industries. He has a proven ability to find ways to save both time and money in any situation. Applying this ability to the field of astrophotography has allowed amateur astronomers across the globe to benefit from his insights into more efficient and cost-effective astronomical imaging techniques.

Kennedy has been using one-shot color imaging equipment since the earliest one-shot color charge-coupled devices (CCD's) became available to the general public. For many years he has been displaying high-quality, color images of deep-space objects for friends and family on his personal website. Through his business website, www.digitalspaceimages.com, he has published and sold his astrophotography images all over the world.

Acknowledgements

This book would not have been possible without the complete understanding, never-ending patience, and full support of my wonderful wife. Since first becoming interested in astronomy, my journey has entailed a lot of long nights out under the stars and a lot of late mornings while I caught up on sleep. My wife has had to endure many hours of taking care of our children while I pursued my hobby, many nights of being woken up when I finally dragged myself to bed, and many hours of hearing me talk incessantly about all aspects of my astronomical pursuits in which she was only mildly interested. Without her encouragement and blessing, I would never have been able to pursue my passion for astronomy and astronomical imaging. While I am grateful for the opportunity she has provided me to pursue this exacting and exciting hobby, I am unbelievably privileged and overly blessed for the love, enjoyment, and fulfillment she has brought to every aspect of my life!

I also could not have written this book without the support and encouragement of my three children who have always taken an interest in what I was doing. We have shared many great moments in front of the eyepiece, and they have marveled time and again at the images I have taken. My first taste of sharing my knowledge of astronomy and astro-imaging with those outside of my family came when my kids asked me to come and speak to their class about my hobby. Year after year, through many presentations to their classes at school, I came to realize that I wanted to help teach other people about the joys of astronomy and astronomical imaging. Above all, I wanted to let people know that with a little time and money, and a lot of dedication and perseverance, anyone can pursue this hobby and doing so will expand their enjoyment of the wonders in the night sky.

Lastly, and most importantly, I want to acknowledge and thank the Creator of the Universe for His marvelous and unfathomable handiwork! Without Him nothing would be possible, especially imaging the breathtaking and awe-inspiring creations that He alone has made…

Contents

1 Some Background 1
   What Is a "CCD" Camera? 1
   One-Shot Color Versus Multiple Color Filter Exposures 7
      Multiple Filter Exposures 7
      One-Shot Color Exposures 8
   Evaluating Your Existing Equipment.. 10
      Telescopes 10
      Mounts 11
      Go-To Capability... 13
      Computers 13
      Location, Location, Location 14

2 Choosing the Right Stuff 17
   Telescopes 17
   Mounts 20
   Telescope Cart... 22
   CCD Imagers 22
   Vibration Control Equipment... 25
   Equipment for Polar Alignment.. 27
   Dew Protection.. 28
   Light Pollution Issues 30
   Focal Enhancement... 30
   Light Box 31

3   Your Imaging Session: An Overview 33
    Planning 33
    Equipment Setup 34
    Polar Alignment 35
    Dark Frames 35
    Building the Imaging Train 36
    Locating Targets 36
    Focusing 37
    Framing 37
    Setting Up an Auto-Guider 38
    Imaging 39
    Light Frames 40
    Breaking Down Your Equipment 41
    Image Processing 41

4   The Planning Stages in Detail 43
    Planning Your Targets 44
    Planning Your Images 50
    Tracking Your Plans 56

5   Setting Up Your Telescope and Polar Aligning It 59
    Setting Up Your Equipment 59
    Polar Alignment 64

6   Focusing and Framing 71
    Focusing 71
    Framing 81

7   Calibration 85
    Dark Frames 86
    Light Frames 90
    Bias Frames 95

8   Taking Exposures and Auto-Guiding 99
    Exposure Times 99
    Signal Versus Noise 102
    Saving Your Images 103
    Auto-Guiding 105

9   Histogram Display and Stacking 113
    Histograms 113
    Stack 'Em Up! 119

10 Image Processing and Enhancement 127
    Processing One-Shot Color Images 128

Calibrating One-Shot Color Images... 128
Aligning and Stacking One-Shot Color Images............................... 129
Processing Individual Color Components.. 130
Combining Color Components 132
Adding a Luminance Layer 136
Processing Your Final Color Image 137
Histogram Adjustment.. 138
Non-Linear Histogram Stretches 141
Sharpening 146
Smoothing or Blurring 148
Digital Development... 151
Deconvolution 153
Summary... 156

11   **Displaying Your Images and Other Possibilities** 157
Displaying Your Images.. 157
Narrowband Filter Imaging.. 159
Solar Imaging... 160
Imaging Manmade Objects 161
Building Mosaics 161
Time Lapse Imaging 164
Astrometry and Photometry... 164
Searching for Novae and Supernovae.. 164
Hunting for Asteroids, Comets, and Dwarf Planets....................... 165

Additional Resources and Star Catalogs 167
Resources.. 167
Magazines. 168
Books 168
Websites 168

Glossary 181

Index... 189

Captions
for Images/Figures

Fig. 1.1 The Lunar Planetary Imager (LPI) from Meade Instruments............ 2
Fig. 1.2 The Digital Sky Imager II (DSI-2) from Meade Instruments............ 3
Fig. 1.3 An image of galaxy M81 showing faint spiral arms 4
Fig. 1.4 The Whirlpool Galaxy (M51).. 6
Fig. 1.5 A Bayer array of color filtered pixels in an alternating
pattern (Courtesy Colin M. L. Burnett).. 9
Fig. 1.6 A luminance image of the Crescent Nebula (C27)......................... 10
Fig. 1.7 A Dobsonian telescope on a platform base,
as opposed to a tripod.. 12

Fig. 2.1 A diagram of an apochromatic (APO)
telescope (Courtesy Tamasflex)... 18
Fig. 2.2 A Schmidt-Cassegrain telescope (SCT) 19
Fig. 2.3 A wedge platform that helps you get your telescope
into polar alignment... 21
Fig. 2.4 An SCT telescope with a wedge platform sitting
on a three-wheeled 'scope cart 23
Fig. 2.5 Vibration control pads used to keep your telescope
from picking up vibrations during imaging....................................... 26
Fig. 2.6 A windscreen can be used to prevent the wind
from causing any vibrations of your telescope,
which are magnified greatly onto your images.................................. 26
Fig. 2.7 A bubble level is used to level your telescope's
base during setup 28

Fig. 2.8   A dew shield extension for your telescope
              will help prevent the onset of condensation on the lens   29
Fig. 2.9   An electric dew protector wraps around your telescope
              and eyepiece lenses to heat them slightly,
              thereby preventing dew from forming   29
Fig. 2.10 Hartmann masks of different sizes and geometric shapes
              can be used to get your telescope to near focus   31
Fig. 2.11 A custom-made light box used for image calibration
              (Courtesy www.digitalspaceimages.com)   32

Fig. 3.1   A dark frame showing "hot" pixels that need
              to be adjusted in your images   36
Fig. 3.2   An auto-guiding telescope and CCD camera "piggybacked"
              onto the main telescope will be used to keep your imaging
              subject on the same pixels of the imaging chip   38
Fig. 3.3   A light frame showing dust motes, smudges, and other
              obstructions in the imaging path that need to be adjusted
              for in your final images   40

Fig. 4.1   The Moon – featuring Tranquility Base where Neil
              Armstrong took his "giant leap for mankind"   45
Fig. 4.2   This image features the Tycho impact crater on the Moon.
              You can get a feel for just how hard the impact was based
              on the white lines of debris that spread out from the crater   45
Fig. 4.3   Saturn – the ringed planet favorite   46
Fig. 4.4   Jupiter – showing the great red spot and the four
              Galilean moons   46
Fig. 4.5   The Dumbbell Nebula (Messier 27)   47
Fig. 4.6   The Blinking Nebula (Caldwell 15)   48
Fig. 4.7   Caldwell 30 is part of a group of interacting galaxies.
              Several other galaxies can also be seen in the background
              of this image (above C30)   49
Fig. 4.8   Light travels through an SCT, bouncing from the primary
              mirror to the secondary mirror and is then focused
              onto the CCD chip   51
Fig. 4.9   A focal reducer is used to increase the field of view
              for a given telescope/imager combination   53
Fig. 4.10 A Barlow lens is used to decrease the field of view
              for a given telescope/imager combination   54
Fig. 4.11 The Ring Nebula (M57) at an image scale
              of 2.11 arc-seconds per pixel   55
Fig. 4.12 The Ring Nebula at an image scale of 0.35 arc-seconds
              per pixel   56

Fig. 5.1 A dropdown rod attachment for your 'scope cart
provides exact precision in setting up
your equipment for each imaging session 61

Fig. 5.2 A CCD imager connected into the eyepiece holder 62

Fig. 5.3 A CCD imager connected directly to an SCT,
in line with the telescope's central axis 63

Fig. 5.4 The altitude (*green*) is one coordinate that determines
how high in the sky your telescope needs to point
in order to locate an object in space. The azimuth
(*red*) is the other coordinate that determines
what compass direction your telescope needs
to point to in order to locate the same object
(Courtesy WikiMedia Commons User: Sigmanexus6) 65

Fig. 6.1 A Hartmann mask can be used to get your telescope
into a rough focus for use with your CCD camera........................... 72

Fig. 6.2 With a Hartmann mask in place, three shapes appear
on your imaging program's display... 74

Fig. 6.3 As you turn the telescope's focusing knob,
the shapes from your Hartmann mask get closer together 74

Fig. 6.4 The closer you get to focus, the closer
the shapes get to each other.. 75

Fig. 6.5 Finally the shapes start to merge together 75

Fig. 6.6 As the light coming through the telescope is focused
onto one spot, you begin to see only the star
on your image display 76

Fig. 6.7 Your rough focus is finished when the star is crisp
and the diffraction spikes from your mask are as long
as you can get them 77

Fig. 6.8 A diffraction mask can be used on the end of your telescope
to give the stars in your images the appearance
of having diffraction spikes 77

Fig. 6.9 The critical focus zone is a small range in the path
of light where the light from all areas of the telescope
lens converge into one spot.. 78

Fig. 6.10 With a rough focus, some stars may still exhibit
a donut-like appearance... 79

Fig. 6.11 With the brighter stars focused, fainter stars
will start to appear; use these smaller stars
to fine-tune your focusing.. 80

Fig. 6.12 As you get closer to the critical focus zone,
you will start to see more and more,
even fainter, stars in your imaging display 80

Fig. 6.13 Using the faintest stars possible for your focus routine
will provide the best possible focus for your images 81

Fig. 7.1 A 15-s dark frame shows up with quite a bit of thermal noise 86
Fig. 7.2 A 60-s dark frame, however, shows even more thermal noise 87
Fig. 7.3 Your image processing program will allow
you to apply calibration frames as part
of the processing procedure 89
Fig. 7.4 A 5-mile wide crater on the Moon can be easily
seen when imaged through a telescope 90
Fig. 7.5 The Crab Nebula (M1) only takes up 420 arc-seconds
of sky as seen from Earth 91
Fig. 7.6 Through your imager, a dust mote appears to take up
a huge swath of the sky 91
Fig. 7.7 A light frame has a flat, even field except where shadows
from dust, dirt, fingerprints, or other obstructions appear................. 92
Fig. 7.8 A light box enables you to take flat-field images immediately
following the imaging of a specific target so you can make
configuration changes when you move onto the next target.............. 94

Fig. 8.1 You will want to take multiple images in case one or more
of your images is ruined by a passing plane or satellite.................... 101
Fig. 8.2 Your imaging control program will offer a variety
of formats you can save your images in 103
Fig. 8.3 Your imaging control program will offer a variety
of methods for saving your images 104
Fig. 8.4 A smaller telescope can be piggybacked onto
your larger telescope along with another CCD camera
for use in auto-guiding 108
Fig. 8.5 An auto-guider requires several settings and other pieces
of information to be assigned for use in auto-guiding....................... 109
Fig. 8.6 An auto-guider will use the centroid of a guide star
to calculate the corrections required to keep the guide
star centered on the original pixel it started on................................ 111

Fig. 9.1 A histogram is a graph showing the number of pixels
with the same value at each value count across
the whole range of the imaging chip's capacity 114
Fig. 9.2 Using the automatic contrast setting, the Veil Nebula
is too faint to be seen on the computer's image display.................... 115
Fig. 9.3 When you shorten the range of data to be displayed
manually, the Veil Nebula shows up very clearly
on the image display.. 115
Fig. 9.4 The histogram display has two settings, either
of which will control the range of pixels
to be displayed on your computer screen 116
Fig. 9.5 The *black* point slider determines which pixels
will be shown as black on the image display 117

Fig. 9.6   The *white* point determines the pixels that will show
            up as completely white on the image display
            (globular cluster M56) 117
Fig. 9.7   Ideally, the data you want to capture should land
            somewhere between one-third and two-thirds
            of the dynamic range of your CCD camera
            (The Triangulum Galaxy, M33) 118
Fig. 9.8   There are several different mathematical calculations
            that can be used to combine multiple images 121
Fig. 9.9   Example of pixel values from nine images used
            in an Average combine process 122
Fig. 9.10   Example of the same nine pixel values used
            in a Median combine process 123
Fig. 9.11   Example of the same nine pixel values used
            with a Min/Max clip where some of the outlier
            data values are eliminated from the calculation 124
Fig. 9.12   The Sigma clip uses the statistical qualities of the data
            set to remove the outlier data values from the pixel
            value calculation 125

Fig. 10.1   Aligning raw images prior to calibration can leave
            a series of flaws that can no longer be cleaned up
            with your calibration frames 130
Fig. 10.2   The color combination program offers a number
            of adjustments that can be made to balance color
            and enhance the final color image 134
Fig. 10.3   The typical histogram for astronomical images
            has four major areas of interest, as shown here 139
Fig. 10.4   A full scale image spreads all of the image's dynamic
            range across the limited display values on your computer
            screen and usually needs a histogram adjustment 140
Fig. 10.5   A linear histogram stretch allocates the selected pixel
            data equally across the available display values
            on your computer screen 140
Fig. 10.6   When set off center, the midpoint slider allocates
            the pixel data from one side of the histogram across more
            of the available display values on your computer screen 141
Fig. 10.7   A logarithmic non-linear histogram stretch will allocate
            more display levels to the brighter pixel data values 142
Fig. 10.8   Adjusting the contrast value of your image
            display can help highlight specific features
            of your targeted subject 143
Fig. 10.9   Adjusting the brightness value of your images can
            make fine details in your images more or less visible 144

Fig. 10.10 The result of a linear and non-linear histogram
stretch along with contrast and brightness
adjustments (spiral galaxy M74) 144

Fig. 10.11 The Curves processing routine in *Photoshop*™ offers
precise control over the shape and function of non-linear
histogram stretches 145

Fig. 10.12 Using the Curves tool, you can assign multiple points
on the curve and adjust each separately to boost only
those parts of the curve where your image
needs enhancement 145

Fig. 10.13 The sharpened image here shows much more detail
than the original image in Fig. 10.11 147

Fig. 10.14 The effects of over-sharpening include artificial details
from sharpened noise and halos around stars in the image 147

Fig. 10.15 This image of the Cocoon Nebula (Caldwell 19) shows
a lot of noise in the image, especially in the background 149

Fig. 10.16 Using the Selection Tools to exclude the stars from the rest
of the image, you can then use a smoothing routine to blur
only the noisy portions of your images 150

Fig. 10.17 Processing images with the Digital Development routine
can save a lot of time and achieve excellent results
with the right image. Compare this image of globular
cluster M13 to the undeveloped version in Fig. 10.11 152

Fig. 10.18 Deconvolution can turn a less than perfect image
into something special 154

Fig. 10.19 Forty-eight iterations using a Lucy-Richardson
deconvolution routine has sharpened the stars
in this image and brought out more of the faint
details within the nebulosity (Iris Nebula, Caldwell 4) 156

Fig. 11.1   An image of the lower left side
of the Eastern Veil Nebula 162

Fig. 11.2   A slightly overlapping image of the upper
right portion of the Eastern Veil Nebula 162

Fig. 11.3   The two overlapping images of the Eastern Veil
(C33) are combined into the start
of a complete mosaic image 163

Chapter 1

Some Background

What Is a "CCD" Camera?

Light is made up of particles known as photons. When you look into the night sky and see the stars, or you see the light from the Sun reflecting off of the Moon or nearby planets, your eyes are actually being struck by photons. Amazingly, these photons were created by nuclear reactions inside of the Sun and other stars and have traveled across vast distances of space and time.

Because our Sun is relatively close in astronomical terms, these photons only have to travel for about 5 min to reach you, perhaps a few minutes longer when they are reflecting off the Moon or the planets of our Solar System. In the case of the other stars, however, these particles of light have traveled millions of miles, some for thousands of years, in order to land on your retinas at that very moment. Think about what that means–most of the starlight, or photons, you see now were created long before you were born. Moving at the speed of light these photons have just now reached Earth where, by chance, after traveling all that time and distance, just happened to land directly on your eyeballs.

When you look through the eyepiece of a telescope, particles of light are entering through the lens and are being focused through the eyepiece into your eye. With each glance, your eye is processing a steady stream of photons that are coming through the telescope. As you observe the light coming through the eyepiece, your eye is constantly absorbing "new" photons coming through the telescope. Since your eye is processing a new set of photons every instant, it has no way to further enhance the light coming through the eyepiece; the newest set of photons coming through the eyepiece is all that your eye and brain can use to "see" the image. But imagine if your eye could gather the light coming through the eyepiece over a

L.A. Kennedy, *One-Shot Color Astronomical Imaging*, Patrick Moore's
Practical Astronomy Series, DOI 10.1007/978-1-4614-3247-0_1,
© Springer Science+Business Media New York 2012

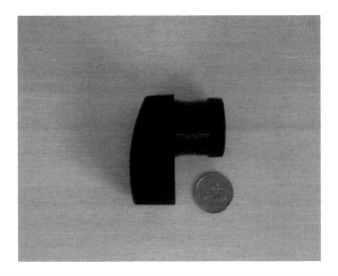

Fig. 1.1 The Lunar Planetary Imager (LPI) from Meade Instruments

period of time and then add it all together. This would enable you to see more clearly the normally very faint objects in the night sky. With the advent of digital imaging, this is exactly what can happen. A CCD camera allows you to gather the stream of photons over time and add them all together in order to better see brighter images of celestial objects.

Specialized digital cameras used for astronomical imaging come in a variety of shapes and sizes. Some very inexpensive models are used primarily to take images of objects in our Solar System, such as the Lunar Planetary Imager from Meade Instruments (Fig. 1.1).

Most models, on the other hand, can be used to capture images not only from the Solar System but also images of astronomical objects well beyond our own galaxy. Some of the most popular and affordable CCD cameras are manufactured by Orion Telescopes and Meade Instruments (Fig. 1.2). Other manufacturers, such as the Santa Barbra Instruments Group (SBIG), produce high-end CCD cameras that can capture very detailed images of night sky objects. These types of high-end imagers have very large sensor chips in them, enabling them to take larger scale, high-resolution images. Typically, the larger the sensor inside the imager, the more the camera will cost.

At the heart of a digital camera is the sensor array, a specially designed computer chip known as a CCD, or charge coupled device. This device is made up of photovoltaic sensors that are highly sensitive to light and more specifically to individual photons. Each chip is made up of hundreds or even thousands of these light sensors lined up in rows and columns. Each sensor, or electrical well, is connected or "coupled" to the sensor on either side of it. The CCD chip works by collecting

Fig. 1.2 The Digital Sky Imager II (DSI-2) from Meade Instruments

and keeping track of the number of individual photons that land on each separate sensor. As photons strike the light sensitive sensors, a tiny electrical charge is created in direct proportion to the number of photons that land on each particular sensor. The more photons that strike the sensor, the bigger the charge gets. After a specified length of time, the amount of charge for each sensor is measured and downloaded as digitized data, which is then used to recreate the image on a computer screen (Fig. 1.3).

Imagine walking outside on an otherwise clear day only to find a menacing-looking rain cloud parked directly over your head. Besides feeling very unlucky, you also know that no one would believe this story unless they could see the cloud for themselves. You would like to take a picture of the cloud, but you don't want to take your camera out in the rain. So instead you line up thousands of buckets side by side in a grid of rows and columns (how you happen to have thousands of buckets in the first place will remain a mystery). After the rain shower, you then count the number of raindrops that have fallen in each bucket. Using that information, you could then recreate an image of the raincloud on your computer based on how much rain fell directly down into each bucket.

Using a grid of squares on your computer that matches the quantity and the location of each bucket, you could color in each square with a shade of gray that corresponds to the amount of rainfall that landed in each bucket. The more rainfall in a bucket, the darker the cloud must have been at that location and the darker gray you would color the square. With buckets that received less rain, you color the square a lighter shade of gray. You would know where the edge of the cloud was because buckets past the edge of the cloud would not have received any rain at all, and they could be colored as clear blue sky. Using this representation of lighter and

Fig. 1.3   An image of galaxy M81 showing faint spiral arms

darker squares you can create a pretty good image of what the rain cloud looked like and the blue sky squares on the grid would define just how big the cloud was.

This is essentially what a CCD camera does. It counts the number of photons that land across the thousands of sensors that make up the imaging chip, just like buckets full of electrical charge. The camera then downloads this information in a digitized format to a specialized graphing program. The graphing program reproduces the information on the screen by displaying the digitized information for each sensor on a separate corresponding pixel in a different shade of gray based on how many photons were collected on that particular sensor. The sensors that received lower amounts of photons are displayed as darker pixels and the sensors that received lots of photons are displayed as brighter pixels. Sensors that received no photons are displayed as black and indicate empty sky (past the edge of the object). This display of lighter and darker pixels line up in the exact sequence of the rows and columns on your camera's imaging chip. Using this graphical representation of how many photons fell on each of the sensors will faithfully reproduce an image of the object the telescope and CCD camera is focused on.

• The CCD, or Charge Coupled Device, gets its name by how it downloads the information into the computer. The program measures the electrical charge in the first sensor in the row and then digitizes the information and downloads to the imaging control program onto your computer. As this information is measured and downloaded, the electrical charges are erased from the first column of sensors and then the charge for each remaining sensor is transferred down the row to the next sensor that it is coupled with. As the process repeats itself, the new electrical charge data for the first sensor in each row is again digitized, downloaded, and erased. The electric charges from the remaining sensors then move down the line again. In a very short time, almost instantaneously, all sensor information is read out in this fashion and the sensors are erased and are ready to start "counting" photons for the next image.

Although it is possible to use a regular digital camera or even a standard film camera, astronomical imaging using a telescope is typically accomplished with a CCD camera. These specialized cameras are much more sensitive to light than regular digital cameras. This sensitivity allows CCD cameras to take much better images of faint night-sky objects. Standard film cameras can be very sensitive to light, and the amount of time that the shutter is open can be varied, which helps them take better images of the night sky than a regular digital camera. The problem with film, however, is that you don't know how your shot came out until you get your film developed. If your picture is out of focus, the subject isn't framed properly, or the picture is saturated from light pollution, you won't know it until it's too late. At this point you have no choice but to start over taking all new photographs. Of course the new set of photos will be subject to the risk of having all of the same issues as the first ones. After taking another set of pictures, if there are problems when you get your new film developed, you will unfortunately have to start all over yet again.

CCD cameras, on the other hand, are usually hooked up to a computer, typically a portable laptop, where the image you take is displayed almost immediately on the computer screen. As you can imagine, this instantaneous view offers an enormous advantage over film cameras. The image focus can be checked immediately, as well as the framing, background saturation, and other aspects of the image. Rather than taking shot after shot, night after night, trying to get everything right, you simply make sure everything looks good before you start taking your pictures. This approach offers you a way to ensure you are going to capture great images in less time and with less money!

Soon, you may even be able to use a tablet computer to do your imaging–which would be the ultimate in convenience. Some tablet computers already come with a standard USB port (needed to connect your CCD camera to a computer). Of course as technology advances, CCD cameras will be manufactured using wireless technology, thereby eliminating the need to use any cables (or USB ports) at all. Some telescopes are already being manufactured with wireless control capability, and CCD cameras won't be far behind. As wireless technology is integrated into the imaging equipment, the use of tablet computers to control the whole imaging process will be an easy choice.

At this point, however, there aren't any "apps" written for imaging control programs, so tablet computers aren't quite ready for use with CCD cameras. Eventually CCD manufacturers, or some enterprising software developers, will begin writing "apps" for CCD imaging. The speed of wireless technology also needs to increase in order to handle the large amounts of data in a timely manner, but these speeds are already increasing by leaps and bounds every year, so it won't be long before all of this technology is ready to be applied to astro-imaging. Of course you can still download and display your images on a tablet computer, just like you can with any

Fig. 1.4 The Whirlpool Galaxy (M51)

other digital device. Always be on the lookout for ways to make your imaging sessions easier, whether it's using smaller, lighter equipment or eliminating as many cables as possible using wireless technology. The less time you spend setting up your equipment, the more time you will have for imaging on those clear nights (Fig. 1.4).

Another new trend in digital imaging that is worth noting is the use of digital video recorders to take shots of objects within our Solar System. These digital video recorders use a type of CCD chip to record images, so using these video recorders for astronomical imaging makes perfect sense. Because these video recorders take multiple images per second, many images can be taken in a short amount of time. As will be covered in a later chapter, these multiple images can be "added" together to come up with a higher-resolution image.

The downside of this approach is that the video frames are images captured in a mere fraction of a second. This does not allow the video recorder to capture the light from fainter objects over time, letting the photons build up before locking in the image. CCD cameras, on the other hand, can take much longer pictures, up to several hours per image, if you have the right equipment and it is set up properly. Not only can CCD cameras take longer images, but like video recorders, they can also take many images that can be "added" together. By taking longer multiple images and then "stacking" them together electronically, you can further enhance and refine your images to reveal an awesome amount of detail. As we'll see in later chapters this process can also reduce flaws, or "noise," in your images.

One-Shot Color Versus Multiple Color Filter Exposures

Multiple Filter Exposures

So, if a CCD camera counts photons and displays the "buckets" in the corresponding rows and columns of pixels on the computer screen only as lighter and darker shades of gray, how do you end up with color images?

There are two ways in which to take color images with a CCD imager. The most common way is to take separate images through colored filters. The other way to get color images is through a specially designed CCD camera that gathers data on all three colors at the same time–the one-shot color imager.

In the first method, a black-and-white (or monochrome) imager is used. Separate images are taken of each targeted object using a series of special colored filters that are placed in the light path before it enters the CCD camera. Although there are many different combinations of colored filters that are used when imaging with this method, the most commonly used are red, blue, and green. After taking separate images through each filter, along with a fourth set of images taken without a filter, these images are combined together in order to form a complete color image.

The major downside of this method is the time involved in taking all of the separate images. The longer an object is exposed for, the more detailed the final image will come out. Many times images are taken for very lengthy periods of time, perhaps an hour or more per image. By using this method of color imaging, each object must be targeted for this length exposure at least four times. In addition, multiple images are usually taken and then "stacked" or added together, in order to increase the amount of detail in your images. If you plan to capture five full-color shots to stack together, you will have to take 20 separate images, five images for each of three colors and five images without a colored filter in the imaging train. At an hour or more per image, you can see how this time really adds up.

Not only does this method require lots of time spent imaging, each image has to be taken accurately so that it can line up with the others when combined to form the whole image. Many times this requires images to be taken on consecutive evenings since the target you are imaging moves across the night sky. Images taken at the zenith will be slightly different, usually more clear and vivid, than images taken closer to the horizon. When trying to image targets that stay close to the horizon, there is usually a limited window of opportunity in which to take the images before they disappear from view.

Taking images on consecutive evenings can also present real problems, depending on how cooperative the weather is in the location where you do your imaging. Out in the desert, where you get great imaging and a clear sky night after night, this is not a big problem, but if you live where the weather is not so accommodating, getting these images on consecutive nights can be a real challenge. There are also other issues associated with the process of taking multiple images through colored filters in order to produce color images. Besides the time it takes to gather four sets of image data, there is also the time it takes to process four different images and then assemble them into one final color image.

Besides taking more time, using this color-imaging method can also be more expensive. The colored filters are specially calibrated pieces of equipment that allow the correct wavelength of light to pass through the appropriately colored filter. The red filter allows wavelengths of light associated in the red spectrum of color to pass through. The blue filter does the same for wavelengths of light associated with the blue color spectrum, as does the green for green wavelengths of light. Like every other piece of equipment used for astronomy, these colored filters are not inexpensive. Additionally, these filters need to be housed in another special piece of equipment called a filter "wheel," which can spin around, allowing you to change the filter for each shot without having to remove the entire imaging train in order to change colored filters. Unfortunately, due to the differing wavelengths of light that each filter allows to pass through, the telescope typically needs to be re-focused for each different colored filter in order to achieve optimal clarity in each set of images. Keep in mind one blurry image in any of the four different image components can lead to a less than desirable result in the final color image after all four sets of images have been combined.

This process is not only the one most widely used by amateur astronomers, it is also the one used by professional astronomers due to its more exacting attainment of data. Of course professional astronomers also have the benefit of having the right imaging location, one that is conducive to multiple exposures in a single night or on consecutive nights. For the amateur astronomer this method is much more difficult and takes a lot more time and effort. Fortunately, there is an easier way to go about capturing color images.

One-Shot Color Exposures

With the advent of one-shot color cameras, a new and easier way to take color images has emerged. As the name implies, all of the colors needed for a full-color image (red, blue, and green) are taken in one shot, as opposed to taking multiple images through different colored filters. This is accomplished through the use of a Bayer array, a specialized set of sensors built into the one-shot color camera's imaging chip. In a Bayer array, the sensors, or pixels, on the imaging chip each have their own colored filter. These colored filters are laid out side by side in an alternating pattern along the rows and columns of pixels so that each pixel captures data in one of the corresponding colors (Fig. 1.5).

As you can see below, the Bayer array contains many more green filtered pixels than it does red or blue. The array was specifically designed this way in order to mimic your vision. The human eye contains specialized light-sensitive cells known as rods and cones. The rod cells are much more sensitive to light in the green spectrum, and your eyes contain many more rod cells than cone cells. Therefore, the Bayer array was designed with more green filtered pixels in the array in order to better reproduce images that you might see if you could view the subject with the naked eye.

Fig. 1.5 A Bayer array of color filtered pixels in an alternating pattern (Courtesy Colin M. L. Burnett)

In order to create the final colored image, a full image component in each color–red, blue, and green–is required. So how do you get a full color image when all you have are these bits and pieces of various colored pixels? Specialized software that comes with the CCD camera mathematically calculates the image data for each color from the various colored pixels into separate color component images for each color. Although the math used in this process can get rather complicated, the basic idea is to take the averages of the data in adjacent pixels and use it to estimate, or interpolate, the data into each pixel of a different color. This process creates a full image in each of the separate colors which is then used to create the final color image. In true one-shot fashion, the same image can also be displayed as an unfiltered, black-and-white image known as a luminance image (Fig. 1.6). These four image components are then combined by the software into the final color image that you will see on your computer screen. In this way a one-shot color CCD camera gives you all of the same data in one-fourth of the time it takes to image using individual colored filters.

There is a small price to pay for this convenience–a minor loss of data resolution. Although the data captured using a one-shot color CCD camera may not be up to the exacting standards of professional astronomers, for the average amateur, these cameras can be used to capture brilliant images of just about any object in the night sky that can be reached by your telescope–far more than can be seen with the naked eye! In addition, with the time saved you can take several sets of images that can be stacked together in order to reduce the signal-to-noise ratio and process them into clearer, cleaner, and crisper images in the end. Also, as the technology for one-shot color CCD imagers advances into the fourth and fifth generation, the quality of images that can be taken are becoming better and better and can easily rival the images captured using some of the older high-end monochrome CCD cameras.

Fig. 1.6 A luminance image of the Crescent Nebula (C27)

Evaluating Your Existing Equipment

If you already have a telescope, eyepieces, and other assorted hardware to go with it, how do you know if these can be used for digital imaging? Well almost any good telescope can be used, but in order to get the most out of your imaging, there are several important features that your telescope must have. If you are just starting out and don't already have equipment, the next chapter will explain what kind of equipment you will need to get started and to take the best digital images that you can for the least amount of money. Of course the better your equipment, the better your images will be. Unfortunately astronomy, and especially astrophotography, is not an inexpensive hobby. There are, however, economical ways to take excellent images without having to buy the most expensive equipment, and of course one easy way you can save money is by choosing a one-shot color CCD camera.

Telescopes

The first thing that you will need to evaluate is your telescope. If you are still using a small department-store telescope, one of those cheap, straight-tube refractor models, where the eyepiece goes in one end and you focus by turning the lens on the other end, this is probably not going to work for you.

Although it would probably be possible to use this type of telescope tube for imaging very bright Solar System objects, there are several issues that would cause this type of telescope to be more trouble than it's worth for most imaging purposes. First, the spindly tripod that these telescopes usually sit on is probably not strong enough to hold your telescope in place with the added weight of the imager placed into the eyepiece. Most of these tripods aren't even strong enough to keep the telescope steady with the slightest hint of a breeze! Secondly, focusing by turning the telescopes lens is not going to give you the precision that is needed in order to take high-quality images. Lastly, using these telescopes can make it very difficult to find targets to image. Without so much as a setting circle (the graduated rings on a telescope that let you know exactly where your telescope is pointing) to find your way around the night sky, finding any objects other than the Moon or the brightest planets will prove all but impossible.

So what kind of telescope *is* good for imaging? Ideally you want something a bit larger – the bigger the better – as you can get higher-resolution images from larger telescopes. A minimum of a 5-in. diameter telescope is required if you want to try to capture decent images of deep-sky objects. Of course an 8-in. diameter telescope would be better, and a 10-in. telescope better still, but of course, the bigger the telescope, the more it will cost.

Fortunately, as the telescope tube's diameter increases, you actually get quite a bang for your buck! This is due to the fact that the important measurement for imaging is not the diameter but the surface area of the telescope lens. A relatively small increase in diameter size equates to a much larger increase in the size of the surface area. Remember the old formula for the area of a circle ($A = \pi r^2$)? It is this squaring factor that provides the notable increase. For instance, a 5-in. diameter telescope offers an area of approximately 20 sq in. ($2.5'' * 2.5'' * \pi = 19.625$ sq. in.), where a 10-in. diameter telescope offers an area closer to 80 sq in. ($5'' * 5'' * \pi = 78.5$ sq. in.). As you can see, doubling the diameter of the telescope's lens actually provides almost four times the available light-gathering surface. That is a bargain!

There are two main types of telescopes, reflectors and refractors. They come in different varieties, however, with some of the more common ones referred to as Dobsonians, Newtonians, and Schmidt-Cassegrains. Excluding the dime-store model starter 'scope, virtually any of these telescopes will work for digital imaging. Of course, the higher-quality telescope you have the better quality images you will be able to take. Each of these models will have its benefits and also some downsides, but each can be suited for astronomical imaging as long as they have a few other necessary features, as you will see below.

Mounts

The next thing you will want to make sure you have for high quality imaging is a good mount for your telescope. It is very important that you have a solid and stable platform from which to image. If your hand shakes while taking a photograph with

Fig. 1.7 A Dobsonian telescope on a platform base, as opposed to a tripod

a standard camera, your pictures will come out blurry no matter how good the camera is. It's the same for digital imaging. An unsteady telescope will produce blurry images, no matter how good the CCD camera is.

Most telescopes sit on a tripod, which holds the tube in place. In order to achieve a steady image, your tripod will need to be sturdy and stable, among other things. If your telescope's tripod doesn't fit the bill, you may want to look at a replacement tripod that is built a bit more solidly. This can't be stressed enough–the more stable that your imaging platform is the better the images you are able to obtain will be. If you use a Dobsonian (Fig. 1.7) or Newtonian telescope (a telescope with a lens arrangement that has the eyepiece in the middle of the tube as opposed to the end), then you are in a little different situation. Instead of being held in a tripod, Dobsonian and Newtonian telescopes are usually attached or rest on a platform-style base that sits on the ground. Here again you will want to make sure you have a solid and stable base for the telescope support. Anything you can do to make the platform more sturdy and steady will dramatically improve the quality of the images you can take.

In order to take images for longer than a few seconds, it is also necessary to have a mount or a base that is designed to move in time with the rotation of Earth. As Earth rotates, the stars seem to "move" across the night sky. If your telescope stays pointed in the same direction while imaging for any length of time, all you will see is a bright streak going across your image. Motorized telescope mounts are specifically designed to compensate for the exact rotational speed of Earth, moving your telescope ever so slowly so that it stays steadily aimed at whatever object you are trying to image. If your telescope mount is not motorized, you may still be able to take very short exposures of bright objects such as the Moon, planets, or star formations, but you will notice your subject moving out of the frame rather quickly, and you will need to continually adjust your telescope's aim in order to keep your targeted object in the imaging frame.

Go-To Capability

Although it is possible to get by without this next feature, to be able to easily find the targets you are interested in imaging it is best to use one of the newer computerized telescopes with a "Go-To" feature built in. The Go-To feature allows you to enter the name or coordinates of any number of different night-sky objects into the telescope's controller. The control program will then automatically move the telescope to point at the right coordinates with the utmost precision.

Keep in mind that you will be able to image far more objects with a CCD camera than you could ever see with the naked eye, so finding these faint targets using the computerized go-to feature makes this task much easier. When your telescope points to a relatively faint object, such as a distant galaxy, you may not be able to see it through your eyepiece, but after the first image is downloaded onto your computer screen, lo and behold, there is the object you are trying to image–right where the star maps said it would be! If you are good at star hopping and finding targets the old-fashioned way, you can always use a telescope that doesn't have this feature, but using a telescope that has the computerized go-to capability will save you a huge amount of time and effort. This is time that can be spent actually taking images rather than trying to find your targets.

Computers

Another important piece of equipment that you will need for astronomical imaging is a computer. Although some of the older model CCD cameras came with their own computing and digital display equipment, with the prevalence of household computers today, CCD camera manufacturers nowadays automatically assume that you will be providing this hardware. The CCD camera will connect to your computer using a standard USB cable, so make sure your computer has at least one USB connection available. Of course this is standard equipment on today's computer models, so if you have a newer model you should be all set. You will also need to make sure that your computer meets the minimum requirements necessary to install and run the programs that drive the CCD camera, but these requirements are usually not too large, so even an older computer will probably suit your needs.

The most convenient type of computer to use for astronomical imaging is a laptop or perhaps even a tablet. Typically you will be taking your equipment outdoors, so the portability of a laptop computer is ideal. Constantly hauling a desktop computer outdoors, or worse yet to a remote imaging location, will prove to be quite cumbersome but it is possible. If you are lucky enough to have your own permanent imaging location such as a shed with a roll-off roof or a semi-permanent clamshell dome, then you can certainly station a desktop computer there and use it for your imaging purposes without having to lug the computer back and forth.

Location, Location, Location

Lastly you will need a good place to image. Ideally you will want a location far from city or suburban lights which can cause your images to be overcome with light pollution, or "skyglow". As will be discussed in a later chapter, there are ways to compensate for this problem, however, so you don't have to feel like you need to drive to a remote location just to take images. A typical suburban backyard can work just fine, but you will want to find a spot that does not have any direct light shining onto your imaging location.

You will also want to find a flat patch of fairly level ground to place your tele-scope on; again, a spot in your backyard or on your driveway will work just fine. It is not recommended that you try imaging from a wooden deck or a balcony, as the tiniest vibrations, such as those caused by walking around, can cause an enor-mous amount of unsteadiness and blur in your imaging. For the same reason you want to make sure you are far away from other sources of vibration, such as an air-conditioning unit or a running motor vehicle. Imaging with a CCD camera requires great precision. Depending on the size of your imaging chip, taking a digital image is the equivalent of trying to take a picture of a dime through your telescope–from 10 miles away! The slightest vibration will be extremely notice-able in your images, so it is best to avoid setting up in these image-ruining locations.

You will need to choose a spot that has an open patch of sky. Obviously things like trees and houses can be a real nuisance when it comes to imaging; you can't image what you don't have a line of sight for. With this said, however, as long as you have at least a small patch of open sky, especially straight over head, you should be fine. Of course the more open area in the location from which you image, the more targets you will have available to take images of. Some interest-ing objects never get very far above the horizon, so having a 360° line of sight is absolutely ideal. Most of the time your imaging location will have some kind of obstruction somewhere, so again, as long as you have a clear patch of sky directly overhead you will be able to find more than enough interesting objects to target throughout the year.

With these basic pieces of equipment—a decent telescope, a steady and stable mount, and a good imaging location—you should have all of the necessary items you need to begin taking incredible images. As you progress in your imaging skills, you will most likely want to find ways to take even better images, using longer exposures. The following chapter will discuss other pieces of equipment that you may eventually want to obtain that can push your imaging to the next level and beyond.

Chapter 2

Choosing the Right Stuff

By now you should have a pretty good idea on whether or not you can use your existing telescope for imaging. Even if your equipment is well-suited for imaging, there are probably other pieces of equipment that you will need to obtain in order to maximize the quality of images you are able to produce. The various pieces of equipment that you may want or need are discussed below. Some of this equipment has uses that will be covered in later chapters of this book. After reading the specific chapter that covers the use of each particular piece of equipment, you can refer back to this chapter in order to determine which items you will want for your imaging needs.

Telescopes

If you don't already have a telescope, or if you find that you need to (or want to) upgrade your existing one, then there are many great models to choose from. One deciding factor on what kind of telescope to purchase is what kind of images you want to take. If you are looking for images that cover wide views of the night sky, then you want to get a smaller refractor type of telescope. These types of telescopes can provide very wide fields of view that are good to use for imaging many of the larger nebulae, various star formations, and other large scale astronomical objects.

An apochromatic (APO) telescope in particular is a great refracting telescope used for this type of imaging (Fig. 2.1). APO's are specifically manufactured to have wide and flat imaging fields that are designed to take outstanding images of very large astronomical objects. If you want to take images of objects that appear

L.A. Kennedy, *One-Shot Color Astronomical Imaging*, Patrick Moore's Practical Astronomy Series, DOI 10.1007/978-1-4614-3247-0_2, © Springer Science+Business Media New York 2012

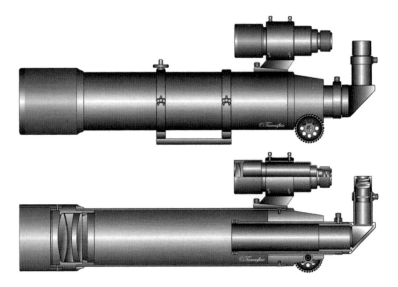

Fig. 2.1 A diagram of an apochromatic (APO) telescope (Courtesy Tamasflex)

smaller in the night sky (based on our vantage point here on Earth), then you will want to get a telescope with a bit more magnifying power.

Although there are many different types of telescopes available with excellent magnifying power, most of which will work just fine for digital imaging, one of the most common types used is known as a Schmidt-Cassegrain telescope, or "SCT" (Fig. 2.2). These compact reflector-type telescopes utilize both a primary and a secondary mirror that increases the effective focal length of the imaging tube. Excellent quality and affordable models are made in a variety of tube diameters by many high-quality telescope manufacturers. These telescopes typically come standard with a high-end, sturdy tripod base, computerized mounts, and more recently the "go-to" programming that will enable an out-of-the-box imaging platform at a reasonable price.

If money isn't an object, then you should look at a Ritchey-Chrétien variety of SCT. These models have excellent imaging qualities in the clarity of both the viewing and the imaging focus that they can achieve and maintain. Their specially designed mirrors offer a higher-resolution imaging solution than a standard SCT. Today's models are also made of new-age materials that minimize the need to refocus them due to variations in the ambient temperature. Although they do cost a bit more than a standard SCT, Ritchey-Chrétien telescopes are probably the best imaging telescopes that money can buy.

For the purpose of imaging most deep-sky objects, it is best to get a telescope with an aperture (lens diameter) of 8 in. at a minimum. Anything smaller than this and you may not be able to capture the desired level of detail in the objects you are trying to image. You may also have trouble finding the objects that you are targeting

Fig. 2.2 A Schmidt-Cassegrain telescope (SCT)

with these smaller telescopes. However, as imaging goes, bigger apertures are not always better. Another important factor to be considered when purchasing a telescope for imaging purposes is the focal ratio.

Focal ratio is defined as the focal length of the telescope divided by the aperture of the primary lens. Focal length is the length of the path that light travels between the primary lens and the eyepiece (or the CCD imaging chip). The aperture is the diameter of the telescope's primary lens. For a typical SCT telescope with an 8-in. primary lens, the focal length is 2,000 mm. The 8-in. diameter of the primary lens converts to a metric measurement of approximately 200 mm. It is important that both figures are converted to the same unit of measurement. Dividing the aperture diameter of 200 mm into the focal length of 2,000 mm provides a focal ratio of 10, typically abbreviated as f10.

The focal ratio of telescopes can vary greatly depending on the type of telescope and the design parameters of the individual telescope manufacturer. Focal ratios are typically categorized as fast and slow. Some telescopes specifically designed for astro-imaging are considered very fast, with a focal ratio of f2, such as the classic 8-in. Fastar telescope made by Celestron (200 mm aperture/400 mm focal length = f2). Any telescopes with a focal ratio of f6 or greater are considered to have a slow focal ratio.

Focal ratio comes into play for astro-imaging in a number of ways, the first of which is ease of use. A telescope with a faster focal ratio is generally easier to use for astro-imaging than one with a slower focal ratio. Because faster telescopes gather light more quickly on the imaging chip you can get the same results in less time. This is a benefit when light pollution is prevalent in your imaging location. It is also a benefit when "seeing" conditions aren't that great. A telescope with a fast focal ratio can image in just about any kind of seeing conditions. The faster light-gathering ability also makes it easier for your mount to track your targeted object, since shorter exposures are able to be used for the same level of data sampling. Telescopes with a slow focal ratio, however, offer more resolution and a higher level of magnification than faster ones do, so there are definitely trade-offs between slower and faster focal ratios.

So how do you decide between a telescope with a fast focal ratio and one with a slow focal ratio? If you are interested in taking images of relatively large objects, such as some of the big nebula clouds, open star clusters, or even constellations, then you are better off with a faster focal ratio telescope. These will generally provide a wider field of view for you to work with. If, on the other hand, you are interested in imaging planets, galaxies, or relatively small-scale deep-sky objects, then you will want to choose a telescope with a slower focal ratio. These will generally offer more magnification power and higher resolution. Just be aware that longer exposures are required with slower telescopes in order to gather the same amount of light onto your imaging chip; longer exposures will require better tracking from your mount.

There are several ways to adjust the effective focal ratio for your telescope, such as by adding a Barlow lens in your imaging train that will make your focal ratio slower. Likewise, adding a focal reducer in your imaging train will make your telescope faster, so don't feel like you are absolutely stuck doing imaging of only one type or size of object with your telescope because of its focal ratio.

As will be explained in a later chapter, you may also want to obtain a smaller telescope and a cheaper, older model CCD imager in order to perform what is called auto-guiding. Although this process is not mandatory, it can improve the quality of your imaging by keeping the telescope pointed to the exact coordinates of the sky, thereby making sure your targeted object remains on the exact same spot of your imaging chip during longer exposures. This second telescope can have a different focal ratio than your imaging telescope, which can give you the opportunity to have one telescope with a slow focal ratio and another with a fast focal ratio. You can then image through the telescope that's appropriate for the object you are trying to take a picture of as well as the imaging conditions on a given evening. A fuller discussion of the types of telescopes and imagers that can be used for auto-guiding will follow when this process is covered later in this book.

Mounts

Once you have selected a good telescope to image through you will need a good mount to set it on. As mentioned in the previous chapter, you will need a mount that is very stable and sturdy.

Fig. 2.3 A wedge platform that helps you get your telescope into polar alignment

When purchasing a new (or used) telescope, especially a larger 10-in. model or bigger, it will typically come with a tripod or platform base. With the larger telescopes, because of the weight of the telescopes themselves, the mounts are usually very sturdy and should be well-suited for imaging purposes. If you are planning to use a smaller telescope, or if the telescope you are looking to purchase does not already come with one, you will want to make sure you invest in a sturdy mount. Again the importance of a sturdy and stable mount cannot be stressed enough. The quality of your imaging will be a direct result of the steadiness of your mount, especially when your telescope is tracking objects for long periods of time across the night sky.

If you are going to be using an SCT-type telescope, another good investment is to buy a "wedge" platform (Fig. 2.3). The wedge sits on top of your tripod, and the base of the telescope attaches directly onto the wedge. The purpose of the wedge is to enable equatorial (polar) alignment, the importance of which will be covered in a later chapter. With today's modern telescopes, having a wedge platform is certainly not critical. Telescopes today and the programs that run them are often designed to compensate for Earth's rotation without having to have a proper polar alignment. Still, in order to get the best-quality images, you will want to be able to polar align your telescope and the wedge is the only way to get the SCT telescope aligned properly.

Telescope Cart

Now that you have a telescope, a sturdy mount, and perhaps a wedge platform, your equipment probably weighs well over 100 lb (45 kg) altogether. Another piece of equipment that you may want to invest in at this point is a cart for your telescope and gear. A 'scope cart is typically a T-shaped metal bar with three inflatable rubber wheels on each point of the "T." Attached to the front wheel is a handle to pull the cart. Your tripod will sit on the metal bar with two legs of your tripod on the cross-bar near the back wheels and the third leg of the tripod on the front bar near the front wheel. The bars have a circular stop in the three spots where the tripod legs sit in order to keep your tripod from slipping off of the bars. The cart also has three small adjustable "feet" that can be raised or lowered in order to level the cart and in turn the telescope, without having to make adjustments to the tripod itself.

Although having a cart for your telescope doesn't have any effect on the quality of the images you can take, it can provide a great benefit to the health of your back. Of course, if you have a permanent arrangement for your telescope, such as a clam-shell dome or a shed with a roll-off roof, this equipment isn't necessary. But if you are like most amateur astronomers, you will be lugging your equipment in and out of your home or garage each time you want to do any observing or imaging. Having your equipment on a cart can not only make the transportation of your equipment easier, but it can also be a great time saver as well (Fig. 2.4).

As long as you are not transporting your equipment to a remote location for imaging but are simply dragging it out to your backyard or driveway, then you can save a lot of time and energy by leaving your telescope set up on your cart in between use. After you are done with an imaging or observing session, you can simply wheel the whole setup into a garage or shed and avoid the time it takes to break down your equipment. When you are ready to use your telescope again, you simply have to wheel the 'scope out of your garage or shed and put it into place on your imaging spot. With a few minor adjustments, your telescope is ready to go. Be sure to cover your telescope with plastic, a sheet, or an old blanket in between use to avoid a buildup of any dirt or dust that can definitely affect your imaging quality.

CCD Imagers

After you have your telescope picked out and you have a sturdy tripod or other support system to place it on, the next major piece of equipment you will need to decide upon is a CCD camera. As was discussed in the previous chapter, there are many different models, sizes, and varieties of imagers available. Most are of the black-and-white or monochrome variety, which require the use of multiple expo-sures through various colored filters. Since you are trying to perform high-quality imaging in less time and spending less money, then you will obviously want to purchase a one-shot color imager.

Fig. 2.4 An SCT telescope with a wedge platform sitting on a three-wheeled 'scope cart

There are quite a few manufacturers of standard monochrome CCD cameras to choose from, but the one-shot color imagers are a newer technology and so there a fewer companies manufacturing these devices. Although there are several new entries in the marketplace, high-quality color imagers have been manufactured for years by Meade Instruments, Orion Telescopes, and Starlight Express. These companies are coming out with bigger and better models every few years, so try to obtain the latest version if you can.

Besides cost, there are a couple of important factors to consider when choosing a CCD camera. First is the size of the sensors on the imaging chip itself. As discussed previously an imaging chip is made up of an array (rows and columns) of light-sensitive sensors, or pixels. Pixel size, as the name implies, is the physical size of each of the individual pixels on your CCD camera's imaging chip. These pixels are typically measured in microns (μ). A micron is the equivalent of one-millionth of a meter – a very small unit of measure indeed. So how do these tiny pixels gather enough light to make a whole image? There is always strength in numbers, and fortunately, an imaging chip is typically made up of hundreds of thousands of these tiny pixels.

Even though small, the size of these individual pixels plays an important role during imaging. The size of the pixels, coupled with your telescope's focal length, will determine the "image scale" at which your equipment will be imaging. Image scale is basically a measurement of how much of the sky each pixel will "see" given the focal length of your telescope. Once again there are trade-offs between ease of use and resolution when it comes to comparing the image scale derived from different camera/telescope combinations. Image scale is measured in arc-seconds per pixel and can be calculated with the following formula:

$$\frac{Pixel\ Size\ (in\ microns) * 206}{Focal\ Length\ (in\ mm)}$$

Ideally you will want your equipment combination to give you an arc-second per pixel measurement in the range of 1–3. Larger pixels will result in a larger arc-second per pixel measurement and vice-versa.

The higher the arc-second per pixel measurement, the easier it is to take images of a given subject for a number of reasons. Higher arc-second per pixel imaging is more forgiving on the accuracy of tracking required to keep your target on the same pixel during long exposures. The larger image scale also gives you a wider field of view, allowing you to capture images of bigger targets. A smaller arc-second per pixel measurement, on the other hand, will give you better resolution and allow you to pick up more detail in your images – *if* you have a clear enough night of "seeing" to use it.

There are limitations to the lower end of the arc-second per pixel range. This is because of the seeing conditions that come from viewing and imaging astronomical objects through Earth's relatively thick and turbulent atmosphere. On a typical night of seeing, you will want to image at an image scale somewhere between 2 and 3 arc-seconds per pixel. On exceptionally clear evenings, you might be able to image around 1 arc-second per pixel for higher resolution and more detailed images. If you are doing planetary imaging, you might even be able to use an image scale even lower.

Whatever you are trying to accomplish with your imaging, you will want to make sure your telescope's focal length and pixel size provide the appropriate range of image scale. Keep in mind that you can always use a Barlow lens or a focal reducer to increase or decrease your effective focal length, hence your image scale, so you do have some flexibility no matter what combination of focal length and pixel size you are imaging through.

The other important factor when choosing a CCD camera is the size of the imaging chip. The imaging chip is made up of an array of pixels in rows and columns. The bigger the pixels and the more rows and columns of pixels there are on your imaging chip, the bigger the chip size will be. All other things being equal, a larger chip size will provide a larger field of view. Bigger chips with more pixels will always capture a wider swath of the sky with any given telescope.

Of course, as with most other items, the larger the size of the chip, the larger the price tag. As a general rule of thumb, you will want to find a versatile CCD camera with an appropriate chip size that fits your budget. No matter what your budget is you can still take great images with chips of any size.

Vibration Control Equipment

As mentioned several times already in this book, even the smallest vibration of your telescope can have a huge impact on the quality of your images. Besides the obvious vibrations that could be caused by machinery (avoid setting up your equipment near air-conditioning units, vehicles, etc.), there are two main ways in which vibrations can occur, causing degradation in your imaging process.

The first way is by movement around your telescope, more specifically, by walking. You would be surprised at how much vibration can be introduced into your imaging system merely by walking around your telescope, even on solid ground or on a concrete or blacktop driveway. Because of the extreme precision and magnification that imaging through a telescope requires, even the smallest amount of vibration will be enormously magnified on your images and can completely ruin your exposures.

Obviously the easiest way to combat this effect is to not walk around your telescope during the imaging process. Get yourself a nice chair and just sit back and relax while your CCD camera and telescope do all the work. In reality, though, this is not always feasible, especially when you are taking long exposures (which are the very images you really don't want to mess up and have to retake).

One excellent way to combat this problem is through the use of "vibration control pads" (Fig. 2.5). As shown here, these are a set of gel-filled plastic and rubber pads about the size of a hockey puck. They come in sets of three and are placed under each leg of your tripod. If you have your tripod resting on a scope cart, then you will want to place the vibration control pads underneath the leveling feet of the 'scope cart prior to adjusting them, in order to level the cart. Be sure to place something on top of the control pad, such as a washer or a coin, in order to prevent the feet of the cart from puncturing the pads.

The second way that vibration can be introduced into your imaging system is by the wind. Although you may not think that a little wind would be able to affect your telescope or imaging system, again you would be surprised as to how little it takes to cause relatively big vibrations and noticeable movement on your imaging chip. Of course you won't be imaging in gale-force winds, but depending on where you plan on doing your imaging, you also may not want to wait for a completely windless night.

The solution for this problem is to have a windscreen of some sort (Fig. 2.6). Although these are generally available commercially, a windscreen can be made very easily with simple materials available at your local hardware store. The windscreen in Fig. 2.6 was made by hanging three tarps along frames made from PVC tubing. When in use, the three sections are fastened together using zip ties. The whole assembly is then set around your telescope in order to block the wind from causing any problems during imaging. Use your imagination and ingenuity when

Fig. 2.5 Vibration control pads used to keep your telescope from picking up vibrations during imaging

Fig. 2.6 A windscreen can be used to prevent the wind from causing any vibrations of your telescope, which are magnified greatly onto your images

designing and building your own windscreen. If you prefer, plans and kits are available at www.digitalspaceimages.com, but you can probably do just as well with your own design.

When deciding on whether to image on a breezy night, keep in mind that wind affects imaging beyond just the vibrations caused by slight movements of your telescope. When imaging through Earth's atmosphere you are actually imaging through several miles of air and moisture. As the wind blows through the atmosphere, light waves get refracted and are bent ever so slightly; this is a phenomenon is known as atmospheric scintillation. The stronger the wind, the more scintillation, or "twinkling," that occurs. The end result of this effect is that you are unable to focus quite as sharply on a windy night as you are on a still night.

Although wind can be a problem, when the alternatives on an otherwise clear night are to do or not do your imaging, many times the best choice is to deal with the wind's effects on your images. By using a windscreen at least you won't have to worry about the additional vibrations that the wind can cause by moving your telescope, thereby making imaging on a windy night that much less difficult.

Equipment for Polar Alignment

As will be discussed in a later chapter, in order to get the best-quality digital images possible, you will want to polar align your telescope. This involves making sure that the central axis of your telescope is perfectly in line with Earth's axis. You can accomplish this by lining your telescope's axis up with Polaris, known as the North Star (for those imagers in the northern hemisphere). Once you have a rough alignment, you will go through a process known as a drift alignment. In order to perform a drift alignment you will need a couple of specialized pieces of equipment.

First you will need what is known as a Barlow lens. This piece of equipment looks like an eyepiece but in reverse. The end that goes closest to the telescope has a lens on that side, while the other end is open. The end with the lens slips into your eyepiece holder, while the other end is used to hold an actual eyepiece.

A Barlow lens is used to increase the magnification of the eyepiece so that your drift alignment process is performed with much higher precision. The more precise you can be with polar alignment, the better tracking your telescope will be able to do and the higher quality images you will be able to take.

Secondly, in order to perform the drift alignment process, you will also need a lighted-reticule eyepiece. This is a special eyepiece that has internal batteries or gets plugged in, usually to a receptacle on your telescope specifically designed for this purpose. When looking through this special eyepiece, you will see a lighted set of crosshairs in the view.

The basic idea of the drift alignment procedure is to center a star or other object in the middle of the crosshairs, then make slight adjustments to your telescope's wedge when you see the star drift out of the crosshairs. The direction in which the drift occurs determines exactly how the wedge needs to be adjusted in order to properly align the central axis of the telescope to Earth's axis.

Fig. 2.7 A bubble level is used to level your telescope's base during setup

(Another piece of equipment that aids in the polar alignment process is called a bubble level. As you can see in Fig. 2.7, this is a small circular device filled with liquid. As the name implies, the level has a bubble inside and also has a small circle at the top of the clear dome. When the bubble is in the middle of the circle, whatever the level is sitting on is perfectly leveled. This device is used to ensure that your telescope's tripod base is level when you set it up. Place the level on your telescope's base and adjust the legs of the tripod or the feet of the 'scope cart until the bubble is inside the circle. Starting with a level base will help you to achieve better polar alignment.

Dew Protection

Depending on the time of year that you are imaging, having dew collect on your telescope can become a real problem. Dew forms when the ambient air temperature reaches the dew point, which is based in part on the amount of moisture in the air. This dew has the effect of fogging up your telescope lens, at which point you are not able to image anything at all through your telescope.

The solution to this problem is twofold. First you will want to have a dew shield for your telescope. This is an extension that gets attached to the lens end of your telescope (Fig. 2.8). The dew shield extends the tube of your telescope past your telescope's lens keeping the outside air from circulating past it. This helps delay the onset of condensation on your telescope lens.

When dew formation becomes a real problem, in order to continue imaging through the night you will need a set of electrical dew protectors (Fig. 2.9). These dew protectors will wrap around the outside of your telescope near the telescope's

Fig. 2.8 A dew shield extension for your telescope will help prevent the onset of condensation on the lens

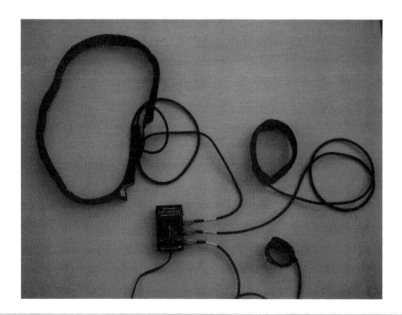

Fig. 2.9 An electric dew protector wraps around your telescope and eyepiece lenses to heat them slightly, thereby preventing dew from forming

lenses and around your eyepiece lens. By providing a gentle amount of heat, the dew protectors keep your telescope and eyepiece lenses warmer than the ambient temperature of the outside air. By keeping the telescope's viewing surfaces at a warmer temperature than the dew point, condensation is prevented from forming on these surfaces. These can be applied at the onset of dew formation, but it will take a little while for them to clear the telescope's lens of any existing condensation, so plan ahead to avoid this problem in the first place.

Light Pollution Issues

Another topic that will be covered in a later chapter is how to deal with the effects of light pollution. One method of dealing with light pollution is through a light pollution suppression (LPS) filter. This is a special filter that goes in your imaging train prior to your imaging chip. The filter is designed to block certain bandwidths of light from passing through it. Streetlights, neon lights, and other artificial lights emit their rays in certain predictable bandwidths. Stars and other natural objects, on the other hand, emit light in a completely different bandwidth of the light spectrum. By blocking that part of the spectrum in which artificial lights shine and allowing all other parts of the spectrum to pass through, a light pollution suppression filter helps to reduce light pollution when imaging.

Focal Enhancement

Focusing will also be covered in much more depth in a later chapter. Most of the focusing process will be done with your telescope's focuser and the visual image on your computer screen. You may also use a computer program – one probably comes with your CCD camera – to improve the quality of your focusing.

A handy tool to get you into a near focus position is known as a Hartmann mask. Again this item can be bought commercially, but it also can be very easily made at home with a few supplies from your local hobby or hardware store. Shown here are several different types of homemade Hartmann masks (Fig. 2.10).

These masks were all made using poster board and a hot glue gun. They can be made out of just about any material – wood, cardboard, etc. – but poster board has the advantage of being lightweight and is more durable than cardboard. The idea is to create a loose-fitting "cap" that can hang off the end of your telescope. In one version of the mask, two crossbars cover a cutout area of the mask. In the other version, geometric shapes are cut into the front of the masks. How these shapes are used to focus the telescope will be discussed in a later chapter.

When making your own version of the masks with the cut-out shapes, two or three geometric shapes should be cut into the surface. The type of the shape is

Fig. 2.10 Hartmann masks of different sizes and geometric shapes can be used to get your telescope to near focus

unimportant; it can be a circle, a square, or even a random shape, but similar-sized triangles seem to work best for several reasons. In the end, as the telescope comes into focus, you won't be able to see the shapes anyway. All of the light coming through the holes in the mask will be focused onto the same spot, and when this happens all that you will see is the round star on which you are focusing.

Light Box

A light box is used in the calibration of images. Calibration involves fixing different types of flaws in your images created by inherent problems with your telescope or CCD camera. This topic will be covered more extensively later in this book. It will point out several different types of calibration that need to be done, including flat-field, or light frame, calibration. One of the most versatile ways to capture the data needed to perform flat-field calibration is through the use of a light box. A light box provides a flat, white surface that is imaged in order to identify any flaws in the optical path such as dust or smudges on the telescope's lens or on any of the other surfaces in your imaging train. This flat-field image is used to "subtract" the flaws from your regular images in order to improve the quality.

Light boxes, for whatever reason, do not seem to be readily available in the marketplace. These boxes take a little more work to build than some of the other equipment presented in this chapter, but with a little handiwork and creativity,

Fig. 2.11 A custom-made light box used for image calibration (Courtesy www.digitalspaceimages. com)

they can still be made at home using relatively inexpensive materials that are readily available at your local hardware or hobby shop, electronics store, and on the Internet. Schematics to build your own light box, as well as assembly kits and even custom-made light boxes (Fig. 2.11), are available at www.digital-spaceimages.com.

Chapter 3

Your Imaging Session: An Overview

Now that you have an idea of all of the equipment you will need in order to perform astronomical imaging, it is time to start talking about the process itself. This chapter will present an overview of what a typical imaging session is like and how to plan for it. Each step in the process will be touched on here and where a more robust explanation is required, those topics will be covered in more depth later.

Planning

As with just about any activity, prior planning can often mean the difference between success and failure; astronomical imaging is no different. The first thing that you will want to spend some time planning for is the weather. Obviously you won't want to be out trying to do imaging in the rain. It won't do you much good to try to image through overcast skies either.

Ideally you will want to image only on very clear evenings and ideally when the sky will be clear all night. If there are just some passing clouds in the sky, you might be able to do some imaging with very short exposures, but don't attempt anything longer than a few seconds. All it takes is for one cloud to pass in front of your target's line of sight and the whole exposure is ruined beyond repair. Windy nights are also troublesome for imaging, so plan to do your best imaging on those clear nights with little or no wind!

L.A. Kennedy, *One-Shot Color Astronomical Imaging*, Patrick Moore's
Practical Astronomy Series, DOI 10.1007/978-1-4614-3247-0_3,
© Springer Science+Business Media New York 2012

Equipment Setup

So you've done your homework. You have found a clear night with no cloud cover and the wind is not whipping around too badly. Knowing this night is in the works, you should also do some research on the objects that you would like to target–ones that will be high in the sky through the course of that evening. With this information in hand, you'll be ready to go outside and begin setting up your equipment for a fruitful night of imaging.

First you'll want to get your telescope outside as soon as possible so the tube (and the air inside) can start acclimating to the ambient temperature; do this perhaps as much as an hour or so before you plan to start your imaging session. This is especially true if your equipment is stored inside, where the temperature is vastly different than the outside air. Depending on the difference between the outside air and your storage location, there could be issues with focusing while your equipment heats up or cools down to match the outside air. You will also want to make sure you get your CCD camera outside as soon as possible for the same reason. The camera needs to acclimate to the temperature as well in order to get accurate calibration images known as "dark frames." As will be discussed shortly, these calibration frames are temperature specific, so you will want to make sure your imager is at the same ambient outdoor temperature before you begin.

If you don't keep your telescope attached to its base or tripod, you will obviously want to get that set up first. Ideally, you will want to set up your telescope on the exact same spot (or as close as you can) every time you image. There are ways to help get your telescope onto the exact spot each and every time; these will be discussed in a later chapter.

Using a bubble level you will want to get your base or tripod as level as possible; this will help you achieve better polar alignment which, in turn, will help you capture better images. Once your telescope is in place, be sure to uncap your tube (this will help speed up the temperature acclimation). If you have a computerized telescope, turn it on and begin the initialization process.

Next you will want to gather all of your other equipment together and make sure it's close at hand. A small folding table is a good place to set your laptop and other equipment on. You will probably also want a comfortable folding chair as you will be spending a fair amount of time in front of your computer during the evening. Don't forget to run an extension cord out to your imaging location. Depending on how much powered equipment you have, you will probably want a power strip of some kind too. There are some good outdoor varieties on the market that are worth every penny. Of course if you are imaging from a remote location, you will need to provide an alternate source of power. There are several good battery-operated power supplies available specifically made for astronomy and astro-imaging. Depending on how much powered equipment you will be running, these battery-operated power supplies can provide enough energy to keep your equipment running through an entire evening of imaging.

Polar Alignment

Once your telescope is set up and acclimated to the ambient temperature, the first thing you will need to do is to get your telescope's axis aligned to the celestial pole. This procedure is known as polar alignment and it is a critical step if you want to take long exposures with your CCD camera. The first step in this process is to get a rough alignment with Polaris, or the North Star (for those imaging in the northern hemisphere). After you have a rough alignment with Polaris, you will want to locate two bright stars, one in the south and one either in the east or the west. You will then use these stars, along with a Barlow lens and a lighted-reticule eyepiece, to perform what is known as a drift alignment procedure. This procedure will be covered in depth during the chapter on polar alignment.

Dark Frames

Another step you will want to perform as soon as possible is taking images known as "dark frames" through your CCD camera. You can probably begin this process while you are in the final stages of your drift alignment procedure. Be sure that your camera has had time to acclimate to the ambient outdoor temperature before you begin as the accuracy of the dark frames for calibration purposes are temperature specific.

CCD cameras are extremely sensitive; in fact this sensitivity is so great that the imager can pick up and display electronic "noise" that gets created by the internal functioning of the camera. This electronic noise is known as "dark current" because it is visible even when no light is reaching the imaging chip. The hotter the temperature the camera is operating at the more dark current gets created. Fortunately, today's CCD imagers are electronically cooled in order to reduce the dark current created by the camera, but this process cannot eliminate the dark current entirely.

A dark frame is an image that is taken to show the dark current that is captured by specific pixels. Dark frame images are taken with the "lens cap on," so theoretically all of the pixels should be reading "zero" or be completely black. In practice, however, you will notice on these images (Fig. 3.1) that some of the pixels are actually gray or even white. These are the specific pixels that are picking up the dark current. Your imaging program will have a special calibration process that will use these dark frames to measure and subtract the value of these pixels from the final data that these pixels record. This allows each pixel to be adjusted, if needed, to contain the true value of the data to be displayed in your final image, without displaying false data created by the dark current.

A separate set of dark frames must be taken for each exposure length at which you are planning to image. These frames should be taken before you start your imaging session and in some cases can be taken the night before or even earlier. After you have your dark frames captured, you will set your imaging control program to automatically adjust these pixels in your data during the imaging process.

Fig. 3.1 A dark frame showing "hot" pixels that need to be adjusted in your images

Building the Imaging Train

Next you will want to connect your CCD imager and any other various pieces of equipment needed for imaging directly onto your telescope. This set of imaging equipment, including the CCD camera, various filters, focal reducers, Barlow lenses, etc., is known as an "imaging train" because all of the equipment is coupled together one in front of the other. Based on your prior planning of the targets that you intend to image, you should have a good idea of what effective focal length and image scale you want to use to capture your images. The desired focal length and image scale will determine what other attachments you will need to use in your imaging train (a Barlow lens, extension tubes, focal reducer, etc.). You may also want to attach a light-pollution filter or any other equipment that you may need for the imaging session.

Locating Targets

After you have your imaging train set up on your telescope, you will want to find your way to the first object that you intend to image. If you have a computerized telescope with "go-to" capabilities, this is just a matter of plugging in the name of the object or the object's sky coordinates. This will at least get you close, probably very close in fact. Some objects that you will want to image are too faint to see right away, but you will be able to take some images with a short exposure length that will let you see the object on your computer screen.

Before you actually lock on to the imaging subject, you will want to pan around a little bit until you find a nice bright star with which to focus on. It is important to do your focusing when you are already near your targeted object. Depending on the type of telescope you are using, your focus could be thrown off by large movements of your telescope. This is a common problem with SCT-type telescopes. The primary mirror can tend to shift slightly with large tube movements, especially when passing the zenith from one direction to another.

Focusing

Once you have a bright star located near your imaging target, you can begin the focusing process. You should start by placing a Hartmann mask on the end of your telescope. You will set your imager to take exposures of about a second or two and begin imaging without saving the images. At this point you will probably see three shapes on the computer screen, depending on what type of Hartmann mask you are using. If you are way out of focus you may only see one of the shapes displayed with the others being off the screen entirely. Not to worry though; as you adjust the focus on your telescope, you will see all three shapes come into view.

As you adjust the focus closer and closer to the right setting, you will see these shapes come closer together until they merge into one. As the shapes merge you should begin making smaller adjustments to your focusing knob; eventually, the shapes will "disappear" and all you will see is the bright star you are focusing on. From here you will want to try to get this star to be as small and as bright as possible on your computer screen.

At this point, you should be able to reduce the length of the exposures you are taking down to a fraction of a second so you can get immediate feedback on your focusing changes. Once you have achieved near-focus, you may also want to consider using a focusing program of some sort to get the best focus possible. The better your focus, the more clear your images will turn out and the more detail you will be able to capture in your images.

Framing

Once your telescope is focused, you will then move back to your targeted object and get it framed properly on the computer screen and the imaging display. Make very small movements of your telescope to accomplish this so you don't lose your subject off of the imaging chip. Keep in mind that some telescopes provide backward or upside-down views, so you will have to play with your movement buttons or knobs to figure out which direction to move the telescope in order to get the subject framed just how you want it. You may also need to increase your exposure length to be able to see parts of the targeted object in order to frame it properly.

Setting Up an Auto-Guider

If you are planning to take long exposures and you have the right equipment in place, it is at this point that you will want to begin getting your auto-guiding equipment prepared. The most common method used to accomplish auto-guiding is by using a second CCD camera attached to a smaller telescope that is piggybacked onto your main imaging scope (Fig. 3.2). The second CCD imager can be a smaller, perhaps older model. The second telescope can also be a smaller, older model if desired, as this equipment will not necessarily be used for imaging. Instead it will be used solely for tracking a bright star and then, using the output from the second CCD camera, to make adjustments to your telescope's positioning in order to keep the guide star on the same pixel of your imaging chip.

You will focus this telescope/imager combination the same way you did for your main 'scope. Using another Hartmann mask made specifically for this size telescope you will get your rough focus manually and then again use a focusing program to get this second telescope focused with fairly good precision. Since you are not imaging with this equipment, your focus doesn't need to be as crystal clear as your main telescope, so you won't need to spend quite as much time on focusing the auto-guiding scope. You will want to achieve decent focusing, however, as the sharper the focus, the more stars you will have to choose from to use as a guide star. Sharper focus also provides for less blur of the guide star, enabling your guiding program to give you better tracking results.

Fig. 3.2 An auto-guiding telescope and CCD camera "piggybacked" onto the main telescope will be used to keep your imaging subject on the same pixels of the imaging chip

Once the auto-guiding telescope is in focus, pick the brightest star visible on the computer screen and select it for your auto-guiding program. An auto-guiding program should be available with your telescope's software if the telescope has the capability to perform auto-guiding. If possible, try not to move your telescope around too much looking for a suitable star to use as a guide star; doing so may mess up the framing that you have already accomplished with your imaging system. Whenever possible, you'll want to use whatever stars are visible on your computer screen as is. Hopefully you'll have some bright ones to choose from (the brighter, the better), but just about any star that you can see on the screen will do. When the star is selected, turn your auto-guiding program on and the auto-guider will begin making fine adjustments to your telescope's pointing position to ensure the guide star (as well as your targeted object for imaging) remains as stationary on the imaging chips as possible.

If your telescope doesn't have auto-guiding capabilities, or if you don't have (or want to) invest in a second telescope and a second CCD imager, you can always try auto-guiding manually. This requires a lighted-reticule eyepiece and an off-axis guiding attachment. The off-axis guider uses a prism to split off part of the light coming through your telescope and then diverts it into the eyepiece, while the remaining light falls on the imaging chip. You will need to center the crosshairs of the lighted-reticule on a bright star within your field of view. You will then need to stare through the eyepiece for the duration of the time you are imaging.

As you notice the guide star moving out of the crosshairs in your eyepiece, you will need to make minute manual adjustments to your telescope's pointing position to re-center the guide star back in the crosshairs of the eyepiece. Before telescopes came equipped with auto-guiding capabilities, this is the way that the guiding process had to be done. It is not an easy task and requires a long time staring through your eyepiece with your attention completely focused on the guide star. Having automated guiding equipment will save a lot of wear and tear on your feet, neck, and back! If this process doesn't sound like something you want to undertake, make sure you spend some quality time polar aligning your telescope and then keep the length of the imaging exposures short enough to match your telescope's ability to track your target.

Imaging

With all of the setup out of the way, you're now ready to begin your imaging session. Set your CCD imaging program to the maximum length of exposures that you can. This will be limited by a number of factors, including your telescope's tracking capabilities, whether or not you are auto-guiding, and the length of time it takes before light pollution and/or your targeted imaging subject begins to saturate your imaging chip. Take as many images as you can of each target as multiple images can be stacked, or added together, to improve image quality and reduce the signal-to-noise ratio.

When you are finished with the exposures for your targeted object, you will want to take more calibration images called "light frames" (see below). You will need to take these calibration frames at the exact same focus that you are imaging at, so it's important to capture the light-frames before you move on to imaging other objects. Once this step is done, you can then move to your next planned object to image, refocus your telescope, set up your auto-guiding equipment again, and start imaging the new object. You can repeat this process as many times as you are able to through the course of the evening.

Light Frames

As stated in the previous section, you will want to take a set of calibration images known as light frames after you are done imaging each of your targeted objects. A light frame is the exact opposite of a dark frame. A dark frame is taken with the "lens cap" on, producing a nearly black image. A light frame is taken against a pure white surface, producing a nearly white image. The purpose of the white frame is to identify any flaws or obstructions in your imaging train, such as dirt, particles of dust, fingerprints, or smudges. At the magnification you will be imaging at, a tiny particle of dust on or near your imaging chip will appear to be a dark smudge or a doughnut shape the size of a large coin on your images.

On the light frame, these flaws will show up as spots that are darker than the rest of white areas on the image (Fig. 3.3). In order to be able to remove these flaws from your images, these obstructions need to appear on the light frame at

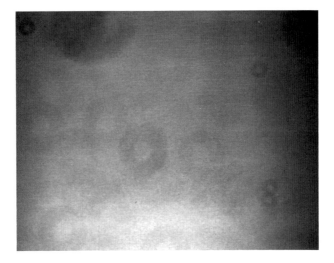

Fig. 3.3 A light frame showing dust motes, smudges, and other obstructions in the imaging path that need to be adjusted for in your final images

the exact same size as they will appear on the images of your targeted subject. This is why it is critical to take these calibration images before you make any adjustments of your telescope's focus or make any big movements of your telescope that could also throw the focus off. As part of the image processing stage, you will use these light frames to remove all of these flaws from your images before you begin any other image-processing routines. Since the pixels that captured these flaws are measurably less bright than the flat white areas of the rest of the image, this data can be used to "add back in" the appropriate values of light to the affected pixels, thereby offsetting the flaws that would otherwise be very noticeable in your images.

Breaking Down Your Equipment

Before you know it you will have finished imaging all of the targets on your list for the evening. Once you have taken your final light frames it will be time to start breaking down your equipment. If you are lucky enough to have a dome or a shed with a roll-off roof, then you don't have much to do. Just put the caps on your telescope lenses and eyepieces and you are all set. If you're like most of us though, you will need to break down your equipment piece by piece. If you haven't already, you may want to invest in a couple of foam-lined carrying cases. These sturdy cases do a nice job of protecting your equipment during transport and storage, and they make things a lot easier to carry once it's all packed up.

If you have your telescope and tripod set up on a 'scope cart, then you simply have to unhook your cables and cords and wheel your 'scope cart into its storage area. If you aren't using a 'scope cart, then you will need to break all of your equipment down and put it away. Be sure you plan some time for this process as it usually takes a bit longer at the end of an imaging session when you are tired and ready for bed.

Once your equipment is put away, you can start processing your images if you want to, but since everything you need is saved on your computer, you will probably want to get some rest and tackle the image processing the next day. You may even want to put off processing your images until a cloudy or rainy night, when you won't be able to go outside and enjoy some time behind the telescope. The data will be waiting for you whenever you're ready.

Image Processing

The last step in the imaging procedure is image processing. As part of this step you will apply any needed calibration frames (light frames, dark frames, or bias frames), stack your images if you have taken multiple exposures, and then perform the final processing to produce your final images.

Entire books have been written about the techniques involved in image processing. It is highly recommended that you pick one or two of these up and familiarize yourself with all of the processing options available. The chapter on image processing that appears later in this book will cover some of the more important steps, especially the critical steps specifically involved in processing one-shot color images. The nice thing about digital images is that once you have your final images processed, if you notice something you want to fix or adjust, you can always go back and tweak the image some more. You are the sole judge of how you want your images to look, so don't be afraid to experiment with different processing techniques to get the images to look just how you want them. When it comes to astronomical imaging…the sky's the limit!

Chapter 4

The Planning
Stages in Detail

Believe it or not, a successful night of imaging starts well before the sky gets dark. Since this book is about saving you time as well as money, the importance of planning your night's imaging session well in advance cannot be understated! Like so many things in life, poor planning leads to poor performance. Astro-imaging is no different; you can save yourself a little time and a lot of headaches by planning your imaging sessions before the Sun goes down!

As mentioned in the previous chapter, one of the first things you will need to plan is a good night for imaging. You can try to plan these evenings by listening to the weather forecast, but the predictions given are usually geared to more general outdoor activities and not specifically to astronomy or astronomical imaging. The weather person can tell you it's going to be a great evening with clear skies and yet big clouds will be passing by overhead, ruining any chance you have of getting the long exposures that lead to great images. What's even more frustrating is the accuracy of the typical weather forecaster's predictions. They are barely worth listening to because their predictions always seem to be hit or miss anyway! But there is a better way to plan your imaging sessions (at least for those of you who do their imaging in North America). A weather forecasting tool specifically designed for astronomical activities is available on the Internet.

At http://cleardarksky.com is an outstanding weather forecasting system, free of charge. Simply go on the website and choose one of the pre-programmed areas that are closest to your imaging location. Save the link into your favorites section and you can go back and check the forecast any time you like. On this site you will find not only information on cloud cover, wind, and humidity for the next 60 h, but also information on several other weather factors that are directly related to the quality of telescope observing and astronomical imaging that you will be able to perform.

L.A. Kennedy, *One-Shot Color Astronomical Imaging*, Patrick Moore's
Practical Astronomy Series, DOI 10.1007/978-1-4614-3247-0_4,
© Springer Science+Business Media New York 2012

The cleardarksky.com website also predicts the transparency of the air (a factor of the water vapor in the atmosphere) and the darkness of the sky (a full Moon can make the sky very bright as it relates to astro-imaging). This website also has a great forecast of a factor called "seeing," which predicts the combination of turbulence and temperature differences at different altitudes. When seeing is good, you will be better able to pick up fine details of celestial objects at high magnification; when seeing is bad this detail may be lost. If you can find an evening when all of the factors line up in your favor, you will have some of the most rewarding imaging time possible!

Planning Your Targets

With this weather forecasting tool set up for easy access, you'll have plenty of time to spend on planning your target list for the next good night for imaging. By looking at the available targets that will present themselves on the nights you have planned for imaging, you can save yourself a lot of time and trouble when it comes to setting up your imaging equipment and performing the actual imaging session. There are many ways to find excellent targets to image. Computerized night-sky planetarium programs are very abundant nowadays. One look at them and, if you're like most people, it will be overwhelming. Before you had only a limited number of objects that could be seen by visual observing through your telescope, but now, given the range and capability of today's CCD imaging devices, the number of targets that you will be able to work with is virtually limitless. So where should you start?

One excellent target for beginners to start with, just as it was when you got your first telescope, is by imaging the Moon (Figs. 4.1 and 4.2). Next you may want to try your hand at the larger planets, Saturn (Fig. 4.3) and Jupiter (Fig. 4.4).

If you want to dive in and try your hand at imaging something a little farther out, try tackling some (or all) of the deep-sky objects that appear on the Messier list. In the late 1700s, French comet hunter Charles Messier developed a list of 110 night-sky objects that were NOT comets. These objects were often mistaken for possible comets when viewed through the existing telescopes of the day. Messier wanted to develop a list of what items to exclude during his comet hunting to avoid wasting time when he ran across them during his searches of the night sky. Ironically, Messier is most famous not for the comets that he discovered but for this list of "non-comets," which includes various star clusters, galaxies, and nebulae (Fig. 4.5).

This list is an excellent place to begin your deep-sky imaging because these objects are among the largest and most easily located deep-sky objects. Most of these objects can be seen in areas of the sky favorable to imaging at some point throughout the year. Information about these objects can be easily found in many books (including this one) or on the Internet. Even better, these items are typically included in the standard databases on telescopes that have a go-to feature.

Fig. 4.1 The Moon – featuring Tranquility Base where Neil Armstrong took his "giant leap for mankind"

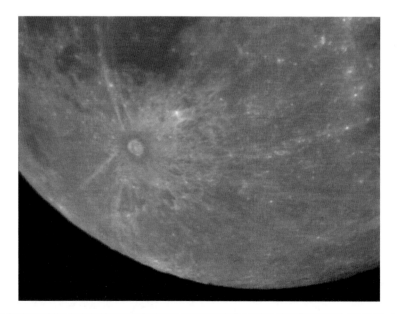

Fig. 4.2 This image features the Tycho impact crater on the Moon. You can get a feel for just how hard the impact was based on the white lines of debris that spread out from the crater

Fig. 4.3 Saturn – the ringed planet favorite

Fig. 4.4 Jupiter – showing the great red spot and the four Galilean moons

Fig. 4.5 The Dumbbell Nebula (Messier 27)

Speaking of which, the go-to feature included with your telescope's database is another great way to find targets to image. They are usually jam-packed with hundreds or thousands of available objects that you can image. You will still need to have an idea of what the items in the database are, so you will know if it is an object that you may be interested in imaging. There are an unlimited number of stars in the databases, but depending on what your area of interest may be, imaging single stars can lose its thrill rather quickly. You will inevitably want to target some of the more interesting subjects like nebulae, star clusters, or even other galaxies.

The Caldwell objects are another well known list of interesting night-sky objects that are excellent subjects for imaging. This list was developed by Sir Patrick Caldwell-Moore (the namesake of this series of books) in the 1950s as a supplement to the list of Messier objects. Messier's list, Sir Patrick Moore noted, was missing several of the night sky's brightest objects. With this in mind, Sir Moore developed a list of 109 equally interesting astronomical objects. This list was developed with the advent of more modern telescopes than Charles Messier had in his day, so the list is naturally much more comprehensive and includes many more interesting night sky objects for viewing and imaging than does Messier's list alone. Coordinates and other information on these objects also appear quite frequently on the Internet and in astronomy-related books (see the end of this book).

As an example, see the bright nebula pictured below (Fig. 4.6), which is known as the Blinking Nebula and is on the Caldwell list as C15. Although this wasn't

Fig. 4.6 The Blinking Nebula (Caldwell 15)

something that Charles Messier was able to find, it is a fairly easy target to image with today's modern telescopes and CCD cameras. Using the Messier and Caldwell lists will enable you to easily find and image hundreds of interesting objects such as the galaxy group known as Caldwell 30 (Fig. 4.7).

Today's modern night-sky planetarium software programs are also an excellent source from which to find suitable objects for astro-imaging. These programs are readily available on the Internet and are often included in CCD camera software programs. By plugging in the location and the date and time of your observing session, whether it's at the current moment, or at some time in the future, these programs can display all of the different night-sky objects that will be available for you to image in the piece of sky that you can see from your imaging location. These programs also list coordinates that you can plug in to your telescope control program in order to direct the 'scope to the appropriate area of the sky. Most planetarium software can also offer useful information about the object being targeted, such as the size, brightness, and what the objects are.

Another great source for imaging ideas is in the astronomy-related magazines that are available in the marketplace. A trip to your local library, or perhaps a year's subscription to one or more of these astronomy-related magazines, can often point out many of the most interesting targets that become visible at different times of the year. Take a look through these magazines, both at the articles and at the pictures

Fig. 4.7 Caldwell 30 is part of a group of interacting galaxies. Several other galaxies can also be seen in the background of this image (above C30)

other imagers have sent in, for your imaging ideas. Even some of the images in the advertisements can often give you interesting ideas about what kinds of objects you would like to image yourself.

Perhaps the most abundant resource for imaging subject ideas is the Internet. A quick search of astronomy, imaging, or astro-photography can reveal hundreds of websites that contain information about, and images of, interesting night-sky objects. It seems most people interested in astronomy as a hobby are very eager to share their knowledge and to help out fellow astronomers, especially new ones (hey, we all started somewhere). The Internet can be a great place to look for help on just about any topic related to astronomy, astro-imaging, telescopes, or anything else for that matter.

No matter where you look for your imaging ideas and target subjects, find something you like, find out when it will be visible at your location, and give it a whirl. With a little luck, your selected target will be showing itself in a favorable position at your location at the current time of the year. As mentioned in an earlier chapter, the best spot in the sky to image at is the zenith – the area directly overhead at your location. Some objects that you might want to image don't ever get to the zenith but instead stay close to the horizon. Still, due to Earth's rotation, there will be times when these objects will rise above the horizon and times when these objects will set below the horizon. Knowing what your exact window of opportunity to image these objects is will help you to plan your imaging session to gain the maximum benefit out of your limited time behind the camera.

Planning Your Images

After having selected an object, or several objects, that you want to image, you should do a little research on these targets. When does the object "rise," when will it be at its best spot in the sky for imaging, what does the object consist of (stars, gas, nebulosity, etc.), how big does the object appear in the night sky? The answers to these questions will enable you to choose the right equipment, imaging time, filters, etc., that you will need to capture the best possible images.

One of the most important pieces of information will be your targets' dimensions – how many arc-seconds or arc-minutes of sky do the objects span. Over the years, astronomers have divided the sky into equal 360° increments around the globe. Each degree is further divided into 60 equal increments known as arc-minutes. Then, as you probably already guessed, each arc-minute is further divided into 60 arc-seconds. With your targets' dimensions in hand, you can determine what accessories you will need to include in your imaging train to get the right-sized field of view. You will want to set up your equipment so that you can fully capture the imaging subject in your camera's frame. In fact, adjusting the field of view to match your targeted subject is such an important aspect of planning your astro-imaging session that it's a good idea to spend some time exploring how it's done.

The field of view, or FOV, is basically the area of the sky that your imaging chip will capture given your specific telescope and imaging equipment combination. The size of the FOV is directly related to several factors, including the focal length of your telescope, the size of the pixels in your imager, the size of your CCD's imaging chip, and the equipment combination's image scale.

In order to ensure you can capture the full view of your imaging target, or perhaps the greatest amount of detail possible, it is important to know what pieces of equipment you can use to alter your field of view. There are several items that can help, including Barlow lenses, focal reducers, and even the use of different-sized telescopes in order to get different fields of view that will best match your targeted imaging subject. So how do all of these factors – focal length, pixel and chip size, and imaging scale – combine to determine the size of your field of view?

As previously discussed, the focal length is how far light must travel between the primary mirror and the focal point (your eyepiece or imaging chip). In a refracting type of telescope, the focal length is pretty much a measurement of the length of the telescope's straight tube. In most other telescopes, mirrors redirect the path

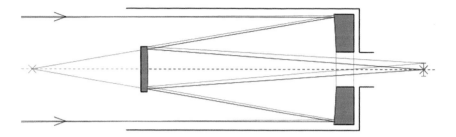

Fig. 4.8 Light travels through an SCT, bouncing from the primary mirror to the secondary mirror and is then focused onto the CCD chip

of light, thereby providing a longer focal length than simply the length of the instrument itself. In a reflecting telescope, such as a Schmidt-Cassegrain, the focal length can be upwards of five times the length of the tube due to the multiple reflections provided by the specially designed mirrors (Fig. 4.8).

If you are unsure about your telescope's focal length, consult your owner's manual or do some research online to find out the specifications for your exact telescope model. The focal length combined with the size of the pixels on your imaging chip will determine the image scale your equipment will be operating at.

Pixel size, as the name implies, is the physical size of each of the pixels on your CCD camera's imaging chip. These pixels are measured in microns (μ), which is the equivalent of one-millionth of a meter. As mentioned before, it is the size of your imager's pixels, coupled with your telescope's focal length, that will determine your image scale. Image scale is a measurement of the amount of the sky that each pixel will be able to "see" when you are imaging. The combined output of the full array of pixels on your imaging chip will, in turn, define your imager's field of view. Image scale, measured in terms of arc-seconds per pixel, can be derived from the following formula:

$$\frac{Pixel\ Size\ (in\ microns)*206}{Focal\ Length\ (in\ mm)}$$

Imaging chips, as we learned in a previous chapter, are made up of hundreds of thousands of pixels. One popular CCD camera, Meade's Deep Sky Imager II, contains an array of 8.6 $\mu \times$ 8.3 μ pixels that is 752 pixels wide by 582 pixels high (437,664 total pixels). The newest version of this imager, the Meade DSI III, contains an array of 6.45 μ square pixels that is 1,360 pixels wide by 1,024 pixels high. This gives the imager a phenomenal 1,392,640 total pixels on this imaging chip's array. It is these rows and columns of pixels that, when added together, make up the dimensions of your imaging chip.

In the examples above, the imaging chip in the Meade DSI II comes out to be 6.5 × 4.8 mm (752 columns of pixels 8.6 μ wide and 582 rows of 8.3 μ pixels high), and the chip in the DSI III comes in at 8.8 mm × 6.6 mm (1,360 columns of 6.45 μ

pixels by 1,024 rows of 6.45 μ pixels). In the end, it is the size of the imaging chip along with the image scale (how much sky each pixel captures) that will determine the size of your field of view. For any given telescope/equipment combination, a larger imaging chip will always provide a bigger field of view.

Continuing with the examples above, let's say you are imaging with the Meade DSI II CCD camera attached to an Orion 80ED telescope. This type of refracting telescope has an 80 mm aperture and a 600 mm focal length. This gives the telescope a fairly slow focal ratio of f/7.5 (600 mm/80 mm = 7.5). Using the image scale formula gives the following result:

$$\frac{Pixel\ size\ (8.6\mu) * 206}{Focal\ Length\ 600\ mm} = 2.95\ arc - seconds\ per\ pixel\ wide$$

$$\frac{Pixel\ size\ (8.3\mu) * 206}{Focal\ Length\ 600\ mm} = 2.85\ arc - seconds\ per\ pixel\ high$$

With this information we can easily calculate the field of view (752 pixels × 2.95 arc-seconds = 2,220 arc-seconds wide; 582 pixels × 2.85 arc-seconds = 1,658 arc-seconds high). Dividing each dimension by 60 in order to convert arc-seconds into arc-minutes gives you a field of view of 37 × 27.6 arc-minutes. If you were to then switch to the Meade DSI III, with its larger imaging chip, the same calculations provide the following image scale:

$$\frac{Pixel\ size\ (6.45\mu) * 206}{Focal\ Length\ 600\ mm} = 2.21\ arc - seconds\ per\ pixel$$

Using this information we again calculate the field of view (1,360 pixels × 2.21 arc-seconds = 3,005 arc-seconds wide; 1,024 pixels × 2.21 arc-seconds = 2,263 arc-seconds high). Converting arc-seconds into arc-minutes puts your new field of view at 50 × 37.7 arc-minutes. As you can see, even though the DSI III has slightly smaller pixels, the larger array provides a much bigger field of view through the same telescope.

Although switching from one imager to another is one way to alter your field of view in order to image objects of different sizes, it is not the easiest or most cost-effective way. Since this book is all about imaging in less time, with less money, let's look at other ways you can alter your field of view without the need for a different CCD imager for each imaging subject. Basically, there are two pieces of equipment that will do the job for far less money. The first is called a focal reducer that, as the name implies, will reduce your telescope's effective focal length, thereby increasing your field of view. The second piece of equipment, which we have discussed previously, is called a Barlow lens. A Barlow lens acts in the opposite way of a focal reducer. It will increase the magnification of the telescope's incoming light, thereby reducing the size of the field of view.

Fig. 4.9 A focal reducer is used to increase the field of view for a given telescope/imager combination

A focal reducer (Fig. 4.9) is a type of small lens that fits into your imaging train. The purpose of the focal reducer is to reduce the effective focal length of your telescope, which also has the added effect of reducing the telescope's focal ratio since focal length is an integral part of the focal ratio calculation.

Typically focal reducers come in two varieties, known as an f/6.3 and an f/3.3. The f/6.3 focal reducer provides a one-third reduction in the effective focal length of your telescope, so in the case of the Orion 80ED telescope, the effective focal length would be reduced from 600 to 400 mm, bringing the f/7.5 focal ratio down to f/4.75 when this focal reducer is in place. Likewise, an f/3.3 focal reducer provides a two-thirds reduction in effective focal length, bringing the Orion 80ED telescope's focal length down to 200 mm and its focal ratio all the way down to f/2.475, a very fast focal ratio.

Using the f/6.3 focal reducer with the Meade DSI III on the Orion 80ED telescope increases your image scale from 2.21 to 3.32 arc-seconds per pixel. Multiplying this image scale by the number of rows and columns on the DSI III's imaging chip brings the field of view from 50×37.7 arc-minutes to 75.3×56.7 arc-minutes, more than double the imaging area. When the f/3.3 focal reducer is used, the image scale increase to 6.6 arc-seconds per pixel and the field of view increases to 150.5×113.4 arc-minutes, large enough to capture almost any deep sky object that you may want to image.

Fig. 4.10 A Barlow lens is used to decrease the field of view for a given telescope/imager combination

So what if you want to image a relatively small object and you would like it to fill the frame of your image in order to show the greatest amount of detail possible? Then, of course, you can go the other way by using a Barlow lens. A Barlow lens (Fig. 4.10) is another type of lens system that provides increased magnification for your optical system. Instead of decreasing the telescope's effective focal length (and its focal ratio), the magnification provided by a Barlow lens increases the telescope's effective focal length. The increase in focal length will reduce the imaging system's field of view accordingly.

A typical Barlow lens will provide twice the magnification power (2×), but Barlow lenses also come in 3× and 4× models, as well. Let's take a look at what happens to the field of view with the same telescope/imaging combination as the example above, when a 2× Barlow is used in the imaging train. Increasing the effective focal length of the Orion 80ED to 1,200 mm provides a very slow focal ratio of f/15 and a still reasonable image scale of 1.1 arc-seconds per pixel.

At this new image scale, the field of view for the Meade DSI III is reduced from 50×37.7 arc-minutes down to 25.1×18.9 arc-minutes, almost a 75% reduction in the imaging area. This smaller imaging area will increase the visual size of your targeted subject on your image, and the smaller imaging scale will provide higher resolution and a greater amount of detail in your images. Keep in mind, however, that the drastically slower focal ratio will now require much longer exposures in order to gather the same amount of light on your imaging chip.

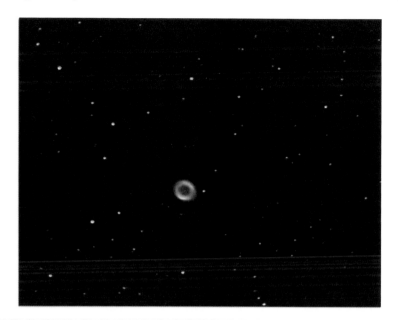

Fig. 4.11 The Ring Nebula (M57) at an image scale of 2.11 arc-seconds per pixel

In order to better understand the effects of altering your image scale, let's take a look at a real example. In Fig. 4.11, an image of M57, the Ring Nebula, was taken with a Meade DSI II imager connected to a 10″ f/10 Schmidt-Cassegrain telescope. In the imaging train a focal reducer was used, which reduced the effective focal ratio to f/3.3. Given this setup and the size of the pixel array on the imaging chip, a field of view 26.9×20.1 arc-minutes was achieved.

Figure 4.12 is the same image taken with a Barlow lens included in the imaging train in place of the focal reducer. The Barlow lens provided twice the magnification power for the telescope and effectively doubled the telescope's natural focal length. Doubling the focal length provided an image scale twice as large as it would be without the Barlow lens. This enabled the same telescope and imager to achieve a field of view that was only 4.4×3.3 arc-minutes. Altering the field of view provided a better way to frame this relatively small subject. As you can see, the smaller field of view made the same object appear much larger in the image and provided for a much greater level of detail as well.

Calculating focal lengths and fields of view can get rather complicated, especially when you have additional equipment attached in your imaging train. Fortunately, there are many resources available on the Internet that can perform these calculations for you. One excellent place to find an image scale calculator is at http://starizona.com/acb/ccd/calc_pixel.aspx. At this website, you simply choose your imager and telescope from dropdown lists (or manually plug in the effective focal length of your telescope), and your image scale and the field of view will be

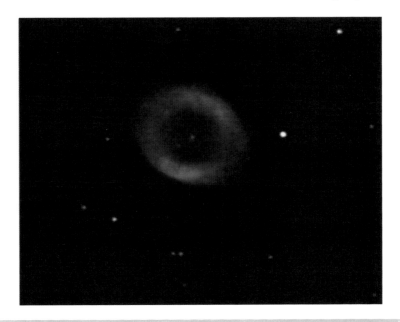

Fig. 4.12 The Ring Nebula at an image scale of 0.35 arc-seconds per pixel

calculated for you. There are other websites available that will automatically calculate your field of view when you are using a Barlow lens, or a focal reducer, to alter your image scale.

As you can see, planning your imaging session and the targets that you want to image on a given night will definitely save you a lot of time when it comes to choosing and setting up your imaging equipment. Knowing in advance just how large your targeted object appears in the sky allows you to select the appropriate equipment needed to alter your image scale, and hence your field of view, to make sure you will be able to properly frame your targeted subject.

Tracking Your Plans

When it comes to the planning process, there is one more thing that will greatly enhance your planning efforts, help you optimize your time spent imaging, and ultimately shorten your learning curve, thereby improving the quality of your imaging. This is the use of a logbook or an imaging journal in order to track all of the information regarding your planning, imaging, and the processing of your images. Although this may not seem necessary, or it may seem like too much trouble, the value of keeping track of all the facts, figures, details, and information regarding your imaging cannot be overstated.

This logbook or journal doesn't need to be anything fancy; a large spiral notebook or two will do just fine. You may be tempted to keep these logs and information on your computer, but there are many times when you will want to reference some piece of information or double check on some procedure and your computer will be tied up with imaging or image processing. Having the information at your fingertips is sometimes a much more efficient and less time-consuming way to go. So what kind of information do you need to keep in this journal?

First you will want to have a section where you keep track of static details. Procedural information that you will need to set up and operate your equipment is a good example. You may also want to write down such things as how to set up your specific equipment; the best ways to put together the imaging train; which way to turn your knobs during setup, polar alignment and focusing adjustments; and any other information that you think will prove helpful to quickly, efficiently, and consistently set up your imaging equipment.

Depending on how frequently you actually image, many of these details can be hard to remember, especially in the beginning or if you have a long spell between imaging sessions. Other information that you may want to keep in this section is data on image scale and field of view calculations for all of your different equipment/telescope combinations. Do the math, or research the figures once, write them down, and then they are there at your fingertips when you go to plan out your imaging and target selection in advance. This can also be a real time saver!

Another section you will want to have in your journal is for information related to your imaging sessions themselves. You should keep track of all of the details of your imaging sessions such as the date and time you start imaging, what kind of equipment you are using in your imaging train, what object(s) you are imaging, the length of the exposures you are using, and how many exposures you take of each object. You should also keep track of details from your planning that relate to your imaging session – what image scale you are imaging at, the size of your field of view, and the size of your targeted object. Finally, you should keep track of the light frames that you take in order to calibrate your images later, along with any other information specific to the imaging session or the imaging subjects.

In the last section of your journal you will want to keep track of all the information pertaining to the processing of your images. As we will discuss in a later chapter, there is a fair amount of work that will need to be done processing your images into their final state. There are many procedural options to choose from during this processing, and you will want to keep track of which choices you made and how they affected your images. You will find that with a little experimentation, you will come across certain procedures that work well for your particular images. Consistency and repeatability, whether in the planning stage, the imaging session, or the processing phase, is the key to shortening your learning curve and speeding you on your way to taking great images in less time, with less money. It is the use of your journal and the detailed information that you record in it that will provide this consistency and repeatability.

Chapter 5

Setting Up Your Telescope and Polar Aligning It

Setting Up Your Equipment

As you begin getting ready for your imaging session, the first step (after the planning phase) will be to set up your equipment. Although this is a pretty straightforward process, there are several options and choices you will have to make at this step. In order to maximize your time imaging, you will need to know how to set up your equipment quickly and correctly so you can take advantage of every opportunity to take the best images possible. This includes not only setting up your telescope in the most advantageous way, but also setting up your CCD camera and the related hardware for your imaging train in the most effective manner.

If possible, you will want to begin setting up your equipment well before sundown. Not only will it be easier to set up your equipment when you can actually see it, but you will save valuable time after the Sun goes down for imaging rather than spending this time fumbling around with your equipment in the dark. After all, the night is only so long, and if you have found a clear night with excellent seeing you will want to capitalize on every minute you can for imaging purposes. The brightest stars (the ones you will use for the polar alignment of your telescope) will be the first to appear in the sky at dusk. Having your telescope and other equipment already set up and ready to go when these stars start appearing allows you to begin your alignment procedures at the earliest possible moment. Once the North Star, Polaris, is visible in the night sky (for those imaging in the northern hemisphere) you can begin your polar alignment routine. If all goes well, you will be able to start imaging within an hour of the Sun going down, assuming the targets that you have selected for imaging are in the right position.

L.A. Kennedy, *One-Shot Color Astronomical Imaging*, Patrick Moore's
Practical Astronomy Series, DOI 10.1007/978-1-4614-3247-0_5,
© Springer Science+Business Media New York 2012

Unless you have a permanent setup in a small dome, or a shed with a roll-off roof, the first thing you will need to decide is where to set up your telescope. Ideally you will want to set it up in the same general location each time you image. If you are imaging in a remote location this may not be feasible, and that's okay. It might take a little longer to go through the polar alignment process, but other than that your setup routine should be the same. If, on the other hand, you are imaging from your driveway or backyard it would be best if you can set up your equipment in the exact same spot every time.

There are a couple of steps you can take in advance to help ensure you are setting your telescope up in the same spot for each imaging session. If you are setting up on a lawn, then you can pour three small circles of concrete where each leg of your tripod should go. These concrete pads only need to be a few inches (8 cm) in diameter, just big enough to fit a vibration control pad on each one. Make sure you sink the concrete down about 6–8 in. (15–20 cm) in order to prevent the pads from being dislodged in any way. Also make sure the top surface is smooth and level; self-leveling concrete will help in this regard. Keep the surface of the concrete pads at ground level so you can run your lawnmower over them without any issues. You will also want to pour these concrete pads in a configuration so that your mount or your wedge will be pointing north when the tripod or 'scope cart legs are set on the pads. This is absolutely necessary for the polar alignment process, so pay close attention to this detail.

If you are setting up your equipment on your driveway, or another level concrete surface, then you can mark the spot that you want your vibration control pads to be set up by painting a circle around each pad with spray paint. As mentioned in a previous chapter, in order to save time setting up your equipment, it would be ideal to have your tripod and telescope set up on a 'scope cart. There is an excellent way to ensure that your telescope will be placed in the exact same spot each and every time, and doing so will drastically cut back on the time it takes to get an accurate polar alignment.

As you can see in Fig. 5.1, an attachment can be added to your 'scope cart that will enable you to set up your equipment in precisely the same spot over and over again. This attachment consists of two rods that drop down from the crossbars of your 'scope cart. These rods will then be inserted into pre-drilled holes in your concrete surface and will allow you to set up your equipment in precisely the same spot over and over again. Since two points define a straight line, using two of these attachments to fix the position of your cart, and therefore your telescope, gives you a level of precision in the location of your imaging spot that can only be rivaled by a permanent setup. These attachments are very easy to make from supplies available at your local hardware store. Plans and kits for these attachments are also available on the www.digitalspaceimages.com website. You can drill the holes in your concrete with a simple masonry drill bit. You will want to make sure you only drill the holes into the concrete by about an inch (2–3 cm) so that you don't penetrate all the way through the concrete slab.

Once your 'scope cart and/or your telescope are positioned, you should use a bubble level or other similar device to make sure your mount is as close to level as

Fig. 5.1 A dropdown rod attachment for your 'scope cart provides exact precision in setting up your equipment for each imaging session

possible. Adjust the legs on top of your vibration control pads to level out the mount. This will help increase the precision of your polar alignment routine, which will in turn improve the quality of your images, especially when taking longer exposures. If you aren't using a 'scope cart to keep your telescope attached to your mount, then you will want to attach the wedge to the tripod and the telescope to the wedge now. In order to save yourself some additional time you may want to consider leaving the wedge attached to your tripod when you break your equipment down. Any little time-saver like this that you can come up with will help you maximize your imaging time.

Next you will want to start preparing the rest of your equipment. Until you run through your telescope's alignment procedure and perform your polar alignment you won't be able to start attaching your imaging train. Depending on how close it is to nightfall and how much you anticipate the temperature to drop when the Sun finally sets, you can probably hook your imager(s) up to the computer and start taking dark frames, if you still need to. Once the stars start appearing in the sky, you can begin the telescope alignment procedure and then using your lighted reticule eyepiece, a Barlow lens, and a high-magnification eyepiece, you can begin the polar alignment process.

After your telescope is polar aligned, you will be able to start building your imaging train. Sometimes you will simply want to attach your CCD camera to the telescope and begin imaging. Most of the time, however, you will want to attach various pieces of equipment in your imaging train, such as a focal reducer, eyepiece extensions, or a Barlow lens, in order to adjust your field of view for imaging. You may also want to consider adding in a special filter known as a Light Pollution Suppression (LPS) filter in order to combat "skyglow" when imaging from a city or suburban location.

Fig. 5.2 A CCD imager connected into the eyepiece holder

The first decision you will need to make is whether you want to set the CCD camera up where your eyepieces are normally attached (Fig. 5.2). With a Newtonian telescope or a straight tube refracting telescope this is your only option. The CCD must be attached in place of an eyepiece, and the light is focused in the normal fashion, just as if you were looking through the telescope with your eye.

With a Schmidt-Cassegrain Telescope (SCT) or a similar model, you have other options. In most cases you can remove the larger fittings where the eyepieces are normally attached and instead attach your camera directly to the telescope along its central axis (Fig. 5.3). This provides the optimal imaging setup for a number of reasons. First, you are able to avoid using light that is reflected off of the internal mirrors that an angled eyepiece holder uses to deflect light from the telescope into the eyepiece. This method circumvents a possible source of misalignment or distortion. Secondly, in this configuration, your auto-guiding telescope (if you are using one) can perform more accurately, making minute adjustments to your imaging telescope in order to keep your targeted subject centered on your imaging chip.

In some configurations, such as if you will be using a Barlow lens in order to magnify your image, then you may have no choice but to attach the CCD camera into the eyepiece holder. Most Barlow lenses are of the eyepiece variety, and this will limit your options. This is perfectly fine, however, and you can still achieve excellent imaging results. If you are not using a Barlow lens or another configuration that requires it, then it is highly recommended that you attach your CCD in the best possible position for imaging – directly through the telescope's central axis and focal point.

Once you have decided which type of setup you want to use, the next decision is to figure out what is the best order of arranging the various pieces of hardware in your imaging train. If you have decided to use one, a light-pollution filter, or any other type of filter for that matter, should go as close to the CCD camera as

Fig. 5.3 A CCD imager connected directly to an SCT, in line with the telescope's central axis

possible. This would be the first item in your imaging train after the CCD camera itself. After the light-pollution filter, you should then attach any eyepiece extension or Barlow lens that you may be using in order to decrease your imaging field of view. The extender or the Barlow lens would then attach directly to your telescope. If instead you are using a focal reducer to increase your imaging field of view, this would be the last item in your imaging train. In this case, the focal reducer attaches directly to the telescope. Once you have your imaging train in place you are ready to hook up your CCD camera(s) to your computer and begin setting up the telescope for imaging.

If you still need to take dark frames, you can put the cap on your telescope and take these dark frames at this point in time. If you already have all of the necessary dark frames taken, then you can begin the focusing process specific to your CCD imager. As you begin the focusing process (see Chapter 6) you will want to use your focusing aids to achieve rough focus and then switch to using a focusing program to finish your focus routine.

If you haven't already you should now attach the dew shield to your telescope even if you know there is no chance of dew forming during your imaging. The dew shield can help prevent any unwanted light from entering your telescope lens and thereby help limit the effects of light pollution to some degree. If there is a threat of dew forming at some point in the evening, then you will want to attach your dew heater to your telescope and imaging train at this point. You don't need to plug the dew heaters in right away, but the act of attaching them can jostle your telescope around and perhaps throw off your alignment if it's not done carefully. Attaching these before you get everything all lined up for imaging only makes sense. You can plug in the heaters when you sense the temperature is dropping close to the dew point or at the first sign of dew beginning to form on your telescope lens.

When you have your CCD imager hooked up to your telescope and especially if you are using a second CCD camera attached to an auto-guiding telescope you will now have a lot of cables hanging off your equipment. It is critical that you do not trip over any of these cables so you may want to find a way to bundle them all together. You can use anything from a zip tie, a bread tie, or even a Velcro strip, to gather the cables together and into one group. If your cables are long enough, you can even run them along the ground and place a mat or a small rug over the cables to ensure you will not trip over them and ruin your telescope's alignment or worse yet, pull any of your equipment off the telescope and possibly damage it.

Polar Alignment

Modern telescopes are very sophisticated pieces of machinery. The manufacturing of the main components hasn't changed all that much over the last few hundred years – they still require a lens on one end and some kind of eyepiece glass on the other, a tube to hold the lenses apart, and maybe some mirrors along the way.

The advent of computers, however, and their application in telescope technology has opened a whole new door for astronomers and even more so for astro-imagers. It used to be a daunting task to just find objects in the night sky, much less take a picture of them. Now it's very simple; all you need to do is to set up your telescope and run through a short procedure so the telescope can align itself properly to the heavens. With the more recent introduction of Global Positioning Satellite (GPS) technology and its incorporation into the telescope alignment procedure, today's telescopes can even find out where they are and what time it is all on their own. Once the telescope is aligned, the computerized go-to technology can point out tens of thousands of objects in its computerized database with pinpoint accuracy and unbelievable precision.

A similar revolution has occurred within astro-imaging for the same reasons. Before computerized telescopes came along, if you wanted to take photographs of night-sky objects you had no choice but to align your telescope to the celestial pole. You also were required to spend a lengthy amount of time hunched over a lighted-reticule eyepiece performing manual auto-guiding of the telescope in order to get a steady enough picture. What used to be a complicated routine and an exacting task has now been replaced by technology. The precision of the telescopes, and the sophistication of the computer programming that's built into them, now allow short-exposure images to be taken without the need for either polar alignment or auto-guiding. The key point here, however, is this only works for short exposures!

Today's computerized telescopes can perform amazing feats in an altitude-azimuth (or "alt-az") mode. In this mode you set the telescope up on a flat base (for an SCT or similar model), as opposed to using a wedge. Through the alignment procedure the telescope will know how far up in the sky to point itself (altitude) and what compass direction to point in (azimuth) in order to find any object in its vast database (Fig. 5.4).

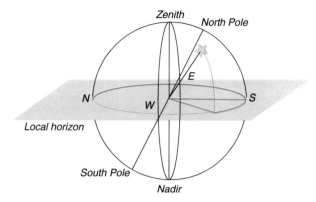

Fig. 5.4 The altitude (*green*) is one coordinate that determines how high in the sky your telescope needs to point in order to locate an object in space. The azimuth (*red*) is the other coordinate that determines what compass direction your telescope needs to point to in order to locate the same object (Courtesy WikiMedia Commons User: Sigmanexus6)

Computerized bases or mounts that come with modern telescopes have gears specifically manufactured to be able to move the telescope at the same rate that Earth rotates. This enables them to keep your telescope pointed at the same spot in the sky all on its own. Without this feature, after a few seconds of exposure your images would start to develop trails when the stars in your image move along the field of view and across the pixels of your imaging chip. When you take short exposures in a telescope's standard altitude-azimuth mode, image processing software can de-rotate the field of view for image alignment. This short-exposure process still limits the amount of detail and the depth of the images you are able to take, however.

Fortunately, there are ways you can overcome the age-old issues that crop up when taking long exposure images. In order to take exposures that are more than a few seconds in length, your mount will need to very accurately track the stars as they "move" across the night sky. Of course what is really happening is the stars are fixed and Earth is rotating, but either way, without pinpoint accuracy in tracking your imaging target, this movement will show up on your images.

The way to get this pinpoint accuracy in your tracking is through the "old-fashioned" process known as polar alignment. In order for your telescope's mount to accurately track your imaging targets – despite the rotation of Earth – it is best if you have your mount aligned with the celestial pole, parallel to Earth's axis of rotation. For telescope users in the northern hemisphere, there is an excellent short-cut to accomplish this feat. Polaris, the North Star, just happens to be almost exactly lined up with the axis of Earth's rotation. Of course, telescope users in Earth's southern hemisphere will need to use a different point in the sky to align to Earth's axis, since the North Star is not visible in the southern hemisphere.

Now here's the tricky part. As stated, the North Star is almost exactly lined up with the axis of Earth's rotation – almost, but not quite! In fact, the celestial pole is about 43 arc-minutes away from the North Star. Although this is a seemingly small distance, it is more than enough to create problems when trying to avoid field rotation in long-exposure images. The better the polar alignment of your mount is, the easier your telescope can track the objects you are imaging. With good polar alignment, you can take unguided images for several minutes, perhaps even up to half an hour, without the dreaded star trails appearing and ruining your images.

The easiest way to polar align a telescope is through a method known as drift alignment. The drift alignment procedure discussed next applies specifically to telescope users in the northern hemisphere. Those in the southern hemisphere can use the same process, but the directions to center stars, the guide star you use, and the direction to make the adjustments will all be different. Although this process seems rather complicated at first, it is fairly straightforward and easy to do. It does require a bit of practice to do efficiently and effectively – practice makes perfect!

In order to accomplish polar alignment your telescope will need to be mounted in an equatorial mode (as opposed to an alt-az mode). To be equatorially mounted a telescope's mount must be able to be set up in such a way that it is parallel to the axis of Earth's rotation. This fact rules out the flat base that comes standard with most SCT type telescopes and will require an equatorial mount known as a "wedge" for these types of telescopes. With a wedge mount, your telescope's flat base is attached to the wedge at an angle. The angle of the wedge can then be adjusted to line up with the celestial pole based on the latitude you are imaging from.

The best way to begin is to make sure the base of your wedge is level. If your wedge has a level built in, adjust the legs or other controls to make sure you start from a level position; otherwise use a bubble level to do the same task. Although not absolutely critical, especially with today's telescopes, it does make the polar alignment process easier and more precise since it minimizes the interaction between the altitude and azimuth settings and the right ascension and declination measurements.

For polar alignment, rather than having the telescope automatically know and adjust for whatever altitude and azimuth is required to find and track your targeted object, you will make adjustments to your wedge, or other mount, in order to ensure that the mount itself (and therefore the axis of your telescope) is set in such a way that the right ascension and declination of the mount and telescope is lined up exactly parallel to the celestial pole.

Right ascension (R.A.) and declination (Dec.) are very similar to altitude and azimuth, but instead of being geographic-based coordinates, R.A. and Dec. are part of a coordinate system for space. Declination refers to an imaginary plane that extends from Earth's equator outward, infinitely, into space (at exactly a 90° angle from Earth's poles). Right ascension relates to an imaginary 360° sphere of coordinates that also extend infinitely into space. These coordinates, measured in "hours" and "minutes" are used to locate the positions of planets, stars, and other objects in the universe from any point on Earth. If you want to learn more about this coordinate system, there is plenty of material available at the library or on the Internet, but it isn't really necessary to know all about these coordinate systems in order to perform polar alignment for your telescope.

To accomplish polar alignment all you will need to do is to make two different adjustments to your mount. The first is the right ascension adjustment, or the exact angle in the sky that your mount should be pointing at to line up with the celestial pole (altitude). The second is the declination adjustment, or the compass direction angle that your mount needs to be pointed to in order for it to be aligned with the celestial pole (azimuth). Using what is known as a drift alignment procedure you will make small adjustments to your mount to ensure that your altitude and azimuth settings give you the correct right ascension and declination positions to put your telescope in line with the celestial pole.

The altitude adjustment, or the angle in the sky your mount needs to point to, is a function of the latitudinal location you are imaging from. Most telescope mounts have an altitude adjustment knob that can get you to the correct general position for this adjustment based on the latitude of your imaging location. If you change your imaging location, you may need to change this latitude setting; fortunately, the altitude adjustment is very forgiving. Chances are, unless you are imaging in a spot that is more than a half-degree of latitude away (about 35 miles or 56 km) from where you last made this adjustment, you will probably not have to readjust this setting.

Once the primary latitude setting is dialed in, you will need to perform the drift alignment procedure to get precise polar alignment. Before you begin this procedure, make sure you run through your computerized telescope's alignment procedure. You will probably need to go into your telescope's database and change the alignment mode to equatorial or polar, as opposed to the standard factory setting of altitude-azimuth alignment. Many imaging software programs now offer ways to perform the drift alignment procedure using your imager to do the alignment. If you will be using this method, attach your imager to your telescope, get it roughly focused, and then follow the procedure manual that comes with your software to perform the alignment.

If you are going to use the more common manual drift alignment procedure then you will need a couple of pieces of equipment attached to your telescope. First you will need to attach a Barlow lens to your eyepiece holder. This provides extra magnification (typically 2×), which will offer a much greater level of precision for your polar alignment. Secondly you will need a lighted-reticule eyepiece attached to your Barlow lens. The lighted-reticule shows up in your eyepiece as a set of red crosshairs that you will use to track the movement of a guide star in the field of view. When you have a star in focus and centered in your eyepiece, you will want to move the telescope very slowly to the left and right and then turn your eyepiece until the star is moving parallel with the lines of the horizontal crosshairs.

Now you are ready to begin the drift alignment procedure, starting with a refinement of the right ascension adjustment (altitude). Pick out a bright star in the east or west, one that is within 5° of the celestial equator. For telescope users in the United States, it would be best to choose a star that is approximately 15–20° above either the eastern or western horizon. You may want to use an astronomical program that will show you the location of the stars in the night sky, the celestial equator, and other astronomical features in order to choose an appropriate star for this procedure.

When you have found an appropriate star, center this star in the crosshairs of the lighted-reticule and then take a look periodically to see if the star "drifts" out of the crosshairs. At this point you are only concerned with drift that occurs to the north or south (up or down); you are not concerned if the star drifts in the eyepiece to the east or west (side to side). What you are trying to accomplish is to make adjustments to your mount so that this star will not drift north or south for a period of at least 5 min. In order to speed up the drift alignment process, don't wait the whole 5 min before checking on the drift. Instead, begin checking the drift after the first minute or so, if the star has already drifted up or down; then you will need to make larger corrections to your mount than you would if the drift occurred later and to a smaller degree.

If you are looking at a star in the east and the star drifts to the north in your eyepiece, you need to adjust your mount's altitude setting slightly downward, if the star has drifted south in your eyepiece, you need to adjust the mount's altitude setting slightly upward. If you are looking at a star in the west, the direction to adjust the mount's altitude setting is reversed; if the star in the west drifts north in the eyepiece, adjust the mount's altitude setting slightly upward, if the star drifts south, then you will want to make the adjustment to the mount's altitude setting slightly downward. The amount of these adjustments will depend on how soon and how far the star has drifted out of the crosshairs of your eyepiece. The sooner (or farther) the star has drifted, the bigger the adjustment you will want to make to your mount. At the beginning of the drift alignment procedure you may want to make an adjustment of approximately two times the field of view of your eyepiece. As the star drifts less and less, you will want to make smaller and smaller adjustments, so that the star you are using only moves by approximately one-half to one-third of your eyepiece's field of view.

After making these slight adjustments, re-center the star and again check the drift after a minute or two. If the star drifts again, make another small adjustment in the appropriate direction. Repeat this process until the star stays centered in the lighted-reticule for at least 5 min. Keeping the star centered for longer provides a more accurate level of alignment, but 5 min without any drift should provide an adequate level of alignment for imaging purposes, especially if you are planning on taking guided images. If you want to take long unguided images, then you may want to perform drift adjustments until the star stays centered for perhaps ten or even 15 min, but keep in mind, every time you make an adjustment, you have to start the countdown over again. You don't want to waste your imaging time getting your telescope to a precise level of alignment that isn't necessary.

Once you have your mount adjusted so that the star on the eastern or western horizon does not drift for a minimum of 5 min, you need to perform the same drift alignment procedure in order to make the necessary refinements to the position of the declination adjustment (azimuth). For this part of the drift alignment procedure, you will need to find a star due south from your location, within one-half of an hour of right ascension from the meridian (the midway point between the east and the west). Ideally you will want to use a star that is slightly to the east of the meridian so that the star moves closer to the meridian as you spend time performing the drift alignment procedure. This star should also be within 5° of the celestial equator; in the United States this would be somewhere around a 40–60° angle above the southern horizon.

Once again, you will want to center this star in the crosshairs of the lighted-reticule eyepiece, making sure the crosshairs are rotated so that when you move your telescope slightly side to side, any movement of the star in the eyepiece is parallel to the lighted crosshairs. During the refinement of the declination adjustment, you are again only concerned with drift that occurs to the north or the south (up or down) in your eyepiece. Once you have the chosen star centered in the crosshairs, begin tracking the time. As soon as you see drift beginning to occur to the north or the south in the eyepiece make an adjustment to your mount as follows. If you see drift to the north, adjust your mount's azimuth adjustment slightly to the east; if you see the star drift to the south, adjust your mount's azimuth setting slightly to the west. The sooner and the larger the drift, the more of an adjustment you will need to make, just as in the previous procedure.

After the adjustment, re-center the star in the crosshairs and begin timing again. As soon as you see drift occur, make another slight adjustment in the appropriate direction and begin the process again. Here, too, the goal is to have the star stay centered in the crosshairs for a period of 5 min without any significant drift. The longer you can keep the star centered in the crosshairs, the more accurate your alignment will be, but 5 min should be accurate enough for imaging purposes, especially if you are planning to use an auto-guider for longer exposures.

Once you have refined the declination adjustment for your mount so that the star in the southern sky stays centered in the eyepiece for at least 5 min, you will then need to go back and perform the right ascension refinement again. Unfortunately, when you make an adjustment on one axis it interacts with the refinement of the other axis. Repeat the right ascension refinement using a star in the east or the west again. You will find that smaller adjustments are needed this time in order to get the star to stay centered for a minimum of 5 min. After this is done you will need to repeat the declination refinement using a star in the sky to the south again. Here, too, you will need to make smaller and smaller adjustments as both axes of your mount get closer to putting your telescope's axis parallel to the rotation of Earth.

If you have a permanent mount, then your telescope should stay reasonably aligned for quite some time and shouldn't need more than an occasional alignment depending on how much your 'scope gets used and the weather, wind, or other factors that could throw your alignment off. If you do not have a permanent mount, then you will have to repeat this alignment procedure each and every time you move the location of the telescope. Of course if you are completely breaking down your telescope, wedge, mount, etc., each time you image, you will have a major task to perform the drift alignment procedure from scratch every time.

In the previous section on equipment setup, several ways were discussed in which you can ensure that you are setting up your telescope, mount, and tripod in the exact same spot each and every time. This can be accomplished a number of ways, such as through the use of concrete pads if you are setting up on a lawn, or through the use of a positioning rod assembly if you are using a 'scope cart to move your telescope to your imaging location.

If you have left your telescope attached to the tripod and mount and you can get your telescope, mount, and tripod set up in the same location each time you go to image, you can cut down the time needed to perform the drift alignment procedure. By positioning your telescope, tripod, and mount in the exact same position you should be able to drastically reduce the amount of time it takes to perform the right ascension refinement procedure. You will still need to perform the declination procedure a couple of times, but even this will go much more smoothly and it will take much less time to perform this procedure. This is yet another reason why having a 'scope cart in order to keep your equipment set up is a great idea.

Although the drift alignment procedure may seem complicated and time consuming, the benefits it will provide to long-exposure imaging is well worth the effort. The first couple of times that you run through it, the procedure will seem like it takes forever. As you get practice doing it, you will get a feel for how big the adjustments are that you need to make for a given amount of drift. Eventually you will gain speed and this procedure will become easier and quicker to do. If you can find a way to keep your tripod, mount, and telescope assembled, and then incorporate a way to ensure you are setting up this assembly in the same spot every time, you will also shave time off of the procedure. Eventually you will be able to get this routine down so that it only takes 20–30 min to get an adequate polar alignment. After you have your telescope in polar alignment, then you are ready to attach your CCD camera to your telescope and begin getting it ready to take images.

Chapter 6

Focusing and Framing

Focusing

With your telescope properly polar aligned you can now attach your CCD imager and the rest of your imaging train to your telescope. The next thing you will need to do is to make sure your camera is precisely focused. It can't be stated enough how important precise focus is to the outcome of your images. Without precise focus, your images will be somewhat fuzzier than they should be, and the fine details you are looking for in your images will be lost.

Focusing during visual observation is very easy – just look through the eyepiece and turn your telescope's focusing knob until you get the clearest view possible and until you are able to see the smallest details about your subject that you can.

Focusing the telescope for use with a CCD imager is not quite as easy. First the field of view is much smaller, so it can be tricky just getting a star to show up on your imaging chip to focus the camera with. Then there is the trick of focusing these stars to the smallest pinpoints possible so that they cover the span of only a few pixels, which we know covers the space of merely a few microns. Fortunately, there are several pieces of equipment you can use to help with the focusing routine in order to get both a rough focus and in turn the precise focus that will be required to take clear, sharp images with your CCD camera.

When you first attach your imaging train to your telescope, the telescope's focus position will be way off from where it needs to be for use with your imager. This is because the CCD imager's resolution is so much greater than even a high magnification eyepiece used through a Barlow lens. Even if you are lucky enough to

L.A. Kennedy, *One-Shot Color Astronomical Imaging*, Patrick Moore's
Practical Astronomy Series, DOI 10.1007/978-1-4614-3247-0_6,
© Springer Science+Business Media New York 2012

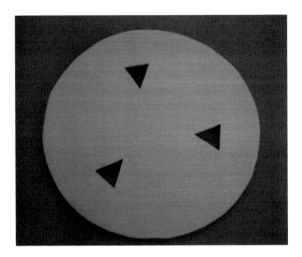

Fig. 6.1 A Hartmann mask can be used to get your telescope into a rough focus for use with your CCD camera

immediately have a star in your imager's field of view, it will be so large and unfocused that you may not be able to see anything more than a bright haze spanning the entire area of your computer screen. For this reason you will typically want to perform the focusing of your telescope in two phases. First you will want to get a rough focus in order to get your telescope's focus adjustment close to the final position. Then after you get your telescope into a rough focus, you can work on getting a very precise focus for your images.

There are several ways to perform rough focusing for your imager, but one of the best and easiest ways is through the use of a Hartmann mask. These are available commercially, and you can find them in a number of sizes at www.digitalspaceimages.com. You can also make one yourself very easily using cardboard, or something a little sturdier such as thick foam-core board. The mask should be fitted to sit on the end of your telescope, and it should have three shapes cut out of the mask (Fig. 6.1). The triangle shape seems to offer the biggest advantages as the shape aids in the focusing process, but in the end it doesn't matter if you use triangles, circles, or any other shapes for your cutouts.

With the Hartmann mask placed on the end of your telescope, locate a bright star close to your imaging target to begin the focusing process. You will want to begin close to your target because any major movement of the telescope, especially as your telescope crosses the meridian, can cause a slight movement of the primary mirror throwing your focus off. This problem is more prevalent on some telescopes than others and tends to be a particular problem with SCT type telescopes. Some of the higher-end SCT models have a "focus lock" that helps alleviate the problem of having the mirror shift when the telescope moves. Make sure you do not over-tighten this lock as it could cause damage or distortion of your primary mirror.

Even with this feature, however, you are still better off focusing on a star that is close to your imaging target; this allows you to avoid the problem altogether. These higher-end model 'scopes also have a motorized focuser that allows you to continue with a more precise focus after the mirror has been locked down with the focus lock.

If your telescope has a motorized focuser, make sure that the focuser is set to the middle of its range when you begin the rough focusing procedure. This will ensure that precise focus can be reached with the motorized focuser after the primary mirror has been locked down with the focus lock, regardless of which direction the focuser needs to be moved.

Even if your telescope does not have a focus lock feature, a motorized focuser is a good piece of equipment to use. Every time you touch the telescope to turn your adjustment knob, you introduce vibration in the system, which plays havoc with the focusing programs you will use to achieve precise focus for your telescope. You may want to consider getting one of the after-market motorized focuser attachments that are available for your telescope.

Using the Hartmann mask, as you get a star within the field of view of your imager, you will see one or more of the shapes that you have cut into your mask appear on the imaging program that is displayed on your computer screen. You will need to see at least two of the shapes in order to know which way to turn your focusing knob in order to bring the image into focus. The good news is that you will always be turning this knob in the same direction when transitioning from an eyepiece to your imager's view. This is an excellent piece of information to write down in your imaging journal and you can refer back to it whenever you begin the focusing routine.

If you can't see at least two of the shapes, move your telescope very slightly in one direction or another until a second shape appears on your imaging display. Now you can begin turning the focusing knob on the telescope until you see the shapes start coming together on your computer screen.

As you begin getting closer to the true focus position you will eventually see all three of the shapes appear on your imaging program. Continue turning your focusing knob in the same direction, and as you get closer and closer to the proper focus, the shapes will begin to come closer and closer together (Figs. 6.2 through 6.6). This is a function of your telescope's lens. A telescope lens is designed so that it gathers all of the light coming through the aperture and then "focuses" all of this light onto one spot – through your eyepiece or onto the imaging chip. So even though the Hartmann mask reduces your aperture to three small shapes that are relatively far apart, the light coming through each shape will eventually all get focused onto the same spot. When you have achieved rough focus, all three shapes will combine into one and your telescope will be in rough focus.

Finish off your rough focusing by making the star appear as crisp as possible. Try to be careful that you do not go too far when turning your focus knob. If you go past the focus point and have to start turning the knob in the other direction you may have trouble with a shifting of the primary mirror. This is a particular problem on Schmidt-Cassegrain and similar types of telescopes. The initial focusing of these types of telescopes is done by moving the primary mirror back and forth on a track that runs along the central axis of the telescope. The mirror is not always securely

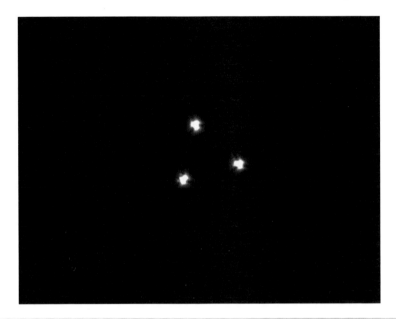

Fig. 6.2 With a Hartmann mask in place, three shapes appear on your imaging program's display

Fig. 6.3 As you turn the telescope's focusing knob, the shapes from your Hartmann mask get closer together

Fig. 6.4 The closer you get to focus, the closer the shapes get to each other

Fig. 6.5 Finally the shapes start to merge together

Fig. 6.6 As the light coming through the telescope is focused onto one spot, you begin to see only the star on your image display

attached on this track and when you start moving the mirror in the opposite direction (by turning the focusing knob the other way) the primary mirror can tend to shift, throwing your focus way off or even pointing the mirror to a slightly different part of the sky. You can see this on your imaging program when the star you are focusing on suddenly moves to a different part of the computer screen or possibly slides off of the screen and out of your field of view entirely.

If this happens, it may be best to go back several turns in the opposite direction and then approach the focus point from the original direction again. This time, try as best as you can not to go past the point of good focus, where you will have to reverse direction yet again. Remember this is only a rough focus, so it doesn't have to be perfect – yet! When you have achieved rough focus, you will lock down the primary mirror (if your telescope has this feature) and then begin the precise focusing routine using your motorized focuser (if you have one), which does not utilize the primary mirror for focusing changes.

When you have achieved a good rough focus, you will also notice that the star on the imaging display has "spikes" coming off of it (Fig. 6.7). This is a function of the focusing mask. In real life, stars are round pinpoints of light as seen through a telescope or on an imaging chip. Because of its aesthetic value, you will often see images of space where stars have these diffraction spikes showing. Many astrophotographers and astro-imagers will use a diffraction mask (Fig. 6.8) to give the stars in their images this appearance, but it's purely an artistic choice, not the way stars really look.

Fig. 6.7 Your rough focus is finished when the star is crisp and the diffraction spikes from your mask are as long as you can get them

Fig. 6.8 A diffraction mask can be used on the end of your telescope to give the stars in your images the appearance of having diffraction spikes

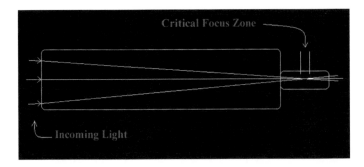

Fig. 6.9 The critical focus zone is a small range in the path of light where the light from all areas of the telescope lens converge into one spot

After you have achieved a rough focus you can start on the precise focus routine. Lock down the focus lock on your telescope, if you have one, and remove the focusing mask from the end of your telescope. You should now be able to see many other stars appear on your imaging program. Instead of being pinpoints of light, some or all of these stars may still appear as little circles or donuts on your imaging display.

Once you have achieved a precise focus, all of the stars should appear as pinpoints of light. The trick to achieving this precise focus is to get your imaging chip into the area of the light path known as the "critical focus zone." The critical focus zone is a very small range in the path of the light coming in through your telescope lens where the light converges into one area (Fig. 6.9). Fortunately the critical focus zone is a range instead of a single point along the path, so there is a little wiggle room to get your imaging chip into this range. The range of the critical focus zone is smaller with faster focal ratios, so getting a precise focus with a fast focal ratio setup is more difficult than with a slower focal ratio setup.

Achieving precise focus just using your eyes to judge is going to be very difficult. Most imaging control programs will come with some sort of focusing software included. It is highly recommended that you use this software to achieve the most precise focusing possible. The better your telescope is focused, the clearer your images will be and the greater the amount of detail your images will show.

If your imaging control program does not include a focusing program, there are separate focusing programs available commercially. Again, it is highly recommended that you acquire and use some kind of focusing program to aid you in your precise focusing routine.

When you begin your precise focusing routine, you will want pick out the faintest star you can see on the screen at this time. The longer the exposure you are taking with your imager during this process, the more faint stars you will be able to see on your imaging program, but the longer the exposure time, the longer it will take to run through the focusing process. You will probably want to take images of around 2 s at this point. This length of exposure will allow you to see fairly faint stars and will also average out any variations in brightness and focus that may occur due to seeing conditions.

Here again seeing conditions are a limiting factor on what kind of imaging results you can achieve. The better the seeing conditions, the easier it will be to focus your equipment and the better focus you will be able to achieve.

Starting with the faintest star you can see on your screen, select this star for your focusing program, typically by dragging a box around the chosen star. If your imaging program offers a zoom feature, switch the view to the maximum amount of zoom available so you can follow along visually while the focusing program tells you whether the focus gets better or worse when you adjust the focuser.

As your imaging program registers the values of the star's focus parameters, you will want to make very minute adjustments to your focusing equipment and see where the new parameters end up. Here is where a motorized focuser really pays for itself. Trying to make very minute focusing adjustments by hand is extremely difficult, although necessary if your telescope doesn't have a motorized focuser or can't be outfitted with an after-market version. Ideally you will be using a motorized focuser in order to make the minute adjustments needed to get your imaging chip into the critical focus zone. Remember, the critical focus zone is small enough to be measured in microns!

As you make these adjustments and get closer to the critical focus zone, you will start to see more, even fainter, stars appear on your imaging program (Figs. 6.10 through 6.13). You will need to turn off the zoom feature of your program, if you are using it, in order to see these stars on other sections of your imaging display. When you begin to see these fainter stars, you should once again select the faintest star that you are able to see. Continue to use the faintest star you can to increase

Fig. 6.10 With a rough focus, some stars may still exhibit a donut-like appearance

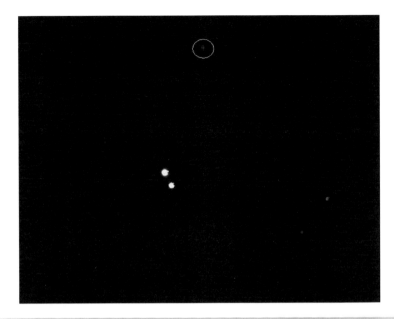

Fig. 6.11 With the brighter stars focused, fainter stars will start to appear; use these smaller stars to fine-tune your focusing

Fig. 6.12 As you get closer to the critical focus zone, you will start to see more and more, even fainter, stars in your imaging display

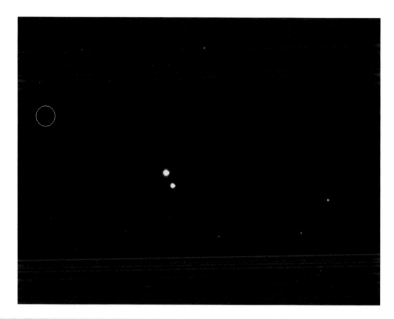

Fig. 6.13 Using the faintest stars possible for your focus routine will provide the best possible focus for your images

your focus quality; the more faint stars that you use for your focusing routine, the more precise your focus will turn out to be.

During the precise focusing routine, in order to ensure you achieve the best possible focus that you can, you will want to actually go right through best focus position and then come back to it. You may even want to do this several times, just to be sure that you are achieving the absolutely best possible focus position.

If you do not have a motorized focuser, than this step could be very problematic, due to the previously mentioned mirror shift that can occur while changing the direction of movement of the primary mirror. If this is the case, you will want to be very judicious in your use of going through the critical focus zone and back. This is another reason why a motorized focuser is the best piece of equipment to have in order to optimize the focusing process. A motorized focuser with a digital readout of the focus position is better yet and will enable you to easily go back and forth through the critical focus zone in order to ensure you have the best possible focus you can achieve.

Framing

Once you have your telescope at the optimal focus position, you are ready to move to your imaging target, get it framed up on your imaging display, and then begin imaging. In order to make it easier to get your subject to appear on your CCD imager's chip it is usually a good idea to make sure the pointing accuracy of your telescope is at the peak of its capabilities.

If you don't have a computerized telescope, then the pointing accuracy of your telescope is in your hands and it's all about your ability to "star-hop" from one reference star to another until you hone in on your target – difficult, but not impossible. In fact, this used to be the only way to locate objects. For the rest of us, thank heavens, there are computerized telescopes that make finding night-sky objects a whole lot easier.

Even with the advantage of a computerized go-to telescope, putting the targeted object on the imaging chip is not always a sure thing. It becomes even more difficult as your magnification increases and your field of view gets smaller.

In order to help you put the target on the chip, you will want to refine the pointing accuracy of your telescope. The best way to do this is to locate several bright stars at different points in the sky, remembering to stay on the same side of the meridian so you don't get any primary mirror shift occurring. Then get these stars centered on the screen of your imaging display and use your telescope's synchronization feature to let the telescope know exactly where it is pointed when the star is on your imaging chip.

After you synchronize several stars in different spots, your telescope should be able to point right back to these stars and bring them onto the imaging chip with no problem. Once fully synchronized, your telescope should be able point to your targeted subject and bring it onto the imaging chip on the first shot. This is very helpful in cases where the imaging subject is very faint and requires many seconds of exposure for the imager to be able to "see" the object. Not having to slew around trying to locate objects that are close, but not dead-on, is a real time saver. Every bit of pointing accuracy that you can achieve with your equipment will pay you back in time spent imaging.

With your equipment polar aligned, your imaging train set up, dark frames in hand (discussed in the next chapter), and the pointing accuracy of your telescope synchronized with the night-sky, you are now ready for the moment of truth. Punch in the coordinates, or select the object you want to image from the database of your computerized telescope, and let the telescope slew to that position.

If you don't have a computerized telescope, then begin your star-hopping from an easily found target, move to the next easily found target, and so on until you are in the right area where your imaging subject is located. With your advance preparation and a little luck the subject will appear on your imaging display. You may also need to increase the exposure length slightly, or modify your display's histogram (discussed in a later chapter), in order to see your intended target.

When you have found your target you will want to ensure that your subject is properly framed on your imaging chip. To do this, you will want to take a short test image – just long enough to see the general features of the object you are imaging in order to make any needed adjustments. With some CCD cameras the display on the imaging control program doesn't reflect the entire breadth of the final images, so just making sure your subject is centered on the imaging display doesn't always give you the ideal placement on your final images.

When you have determined which way and how far your subject needs to be moved in order to get the placement you want, make very small adjustments to your telescope's position in order to move the imaging subject to the location you want it in. It will be helpful to watch a bright star on your imaging display using very short exposure lengths of a fraction of a second while you move the 'scope to place your targeted object in the proper position.

When you think you have your imaging subject appearing in the proper spot, it is a good idea to take another short test exposure just to be sure. Make additional corrections, if needed, so your imaging subject is right where you want it to be; once you start imaging you won't be able to change the placement of the subject without having to start over from scratch. When your target is framed in the exact spot you want it to be, you are ready to begin imaging.

Chapter 7

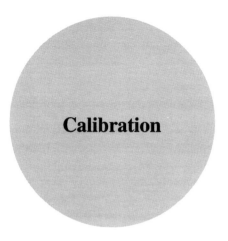

Calibration

Calibration, also known as image reduction, is the process of taking several different types of images in order to compensate for any operational issues or flaws in your CCD camera, telescope lens, or other equipment in your imaging train.

There are two main types of issues or flaws that you will need to compensate for. The first is for operational issues involved with the CCD camera and imaging chip. As mentioned in a previous chapter, CCD cameras are so sensitive that they can pick up readings from the operation of the imager itself and display them as bright dots on your images; these dots are known as hot pixels. To compensate for these you will need to take a series of calibration images known as dark frames.

The other thing that you will need to compensate for is any dust, fingerprints, or smudges that may be present on your telescope's lens, the imaging chip, or any of the other lenses or surfaces that are present in the light path through your imaging train. In order to compensate for these types of flaws, you will need to take another series of calibration images known as light frames.

Dark frame calibration images should be taken before you start imaging, if possible. Your imaging program will use these calibration frames during the imaging procedure in order to compensate for any hot pixels that could otherwise show up in your images. Light frames are taken after you are done imaging your subject(s), and these calibration images are used at the onset of your image processing routine in order to remove any of the flaws mentioned above from your final images.

L.A. Kennedy, *One-Shot Color Astronomical Imaging*, Patrick Moore's
Practical Astronomy Series, DOI 10.1007/978-1-4614-3247-0_7,
© Springer Science+Business Media New York 2012

Dark Frames

The great thing about CCD imagers is their sensitivity to light. The whole idea behind a CCD imager is that it gathers and collects individual photons streaming in through your telescope lens. This sensitivity does come with a price, however. Today's imaging chips are so sensitive that they can also gather photons from sources other than the light coming through your telescope. The main culprit of these unwanted photons is the internal heat generated by the operation of your CCD camera itself. This unwanted heat is known as dark current and the photons created by this heat that show up on your imaging chip are known as thermal noise.

Thermal noise, as the name implies, are photons generated by heat (thermally), and they are called noise because they are not something that you want appearing in your final images. A CCD camera is an electronic device, and like any electronic device the current, or the movement of electrons within its circuitry, generates heat. This heat is radiated out of the internal circuitry of the imaging device as photons, some of which will be picked up by the imaging chip. The longer the exposures you are taking, the hotter your CCD camera will become and the more photons and thermal noise will be generated (Figs. 7.1 and 7.2).

Fig. 7.1 A 15-s dark frame shows up with quite a bit of thermal noise

Fig. 7.2 A 60-s dark frame, however, shows even more thermal noise

In order to minimize this effect, most CCD cameras are artificially cooled. In older models this would entail water-cooling where you actually had a bucket of ice water that connected to your imager through tubing that circulated the cold water around your CCD camera. Fortunately, new technology available in modern imagers uses thermo-electronic cooling (it's all done for you electronically), so you will not have to mess with buckets of water and wet tubing hanging off of your imaging train. Nevertheless, even with thermo-electronic or any other type of cooling, dark current and the resulting thermal noise still occurs. The good news, though, is that this effect is repeatable and therefore predictable, allowing these flaws to be compensated for.

The method of compensating for dark current and thermal noise is through the use of dark frames. A dark frame is basically an image taken with a CCD camera that has its "lens cap" on. With no photons from outside the camera landing on the imaging chip, the only light value readings that will show up on these images are those coming from the dark current. These dark frames are then used to subtract the value of each hot pixel from the value of the corresponding pixel in your images, resulting in a true value of the light gathered by each pixel.

Most imaging control programs include a procedure that will subtract the appropriate dark frame from your images while you are taking them. If not, this calibration procedure can always be performed afterwards.

The main predictor of hot pixels on your imaging chip is the temperature at which your CCD camera is operating. The longer the exposure you are taking through your imager, the hotter your CCD camera will operate and, as logically follows, the more dark current is created. For this reason, a separate set of dark frames must be taken for each length of exposure that you will be using. For example, if you plan on taking exposures at 1 s, 30 s, 5 min, and 30 min in length, you will need to take a set of dark frames of 1 s, another set at 30 s, another set at 5 min, and so on. This way the pixels that are affected at that temperature range will show up as hot pixels in the dark frame and the data will be properly adjusted in your exposures. Additionally, as the ambient temperature changes by more than about 5° or so, a new set of dark frames should be taken in order to account for the changes in the amount of dark current generated.

In order to avoid any extraneous sources, such as a cosmic ray strike, or to avoid any inconsistent patterns in your dark frames, generally about five separate images need to be taken at each exposure length. This set of dark frame images are mathematically averaged by your imaging control program in order to create a single dark frame for each exposure length and temperature reading.

As you can see this process may take a little advance planning – the time it takes to gather five 30-min dark frames can really add up. Fortunately, the dark current at a given exposure and temperature is very consistent, so you can build up a "library" or "catalogue" of dark frames that can be used again and again. Taking dark frame images for a dark frame "library" is a great thing to accomplish on those rainy or cloudy nights when you can't do any imaging anyway.

Imaging control programs will typically have some sort of process to take dark frames designed right in to the program. As mentioned, the process of taking dark frames basically involves taking images with the "lens cap" on or some type of a cover on the CCD camera's lens. If you are taking the dark frames with the CCD imager already attached to your telescope, then you will need to cover your telescope tube either by putting the tube cap in place or covering the end with a completely opaque cloth, a piece of cardboard, or something else that will ensure that no outside light lands on the imaging chip. Any light or bright pixels that still show up on your imaging chip are the result of dark current or flawed pixels in your imager. Most imaging chips have a few flawed pixels that will run "hot" and will always show up as bright even when the temperature of the imager is very cold. These too will be corrected when you calibrate your images using the dark frames.

Oftentimes you may not know exactly how long you will want to take your images for, so it can be useful if you go ahead and take a series of dark frames at various lengths of time. Your imaging program should be able to match the right dark frames with your images depending on how long you set the exposures for during the imaging session. This works fine for images up to about 5 min or so, but because you have to take several dark frames for the full length of time of each exposure length, taking unnecessary dark frames of longer length can be very time consuming. If you plan on taking longer length images then you have to plan on spending a good deal of time taking your dark frames. Here again, taking dark frames of longer exposures is an excellent use of a rainy or cloudy evening.

If your imaging control programs offers the feature of removing dark frames during image capture, this is definitely the route you want to take. Many of these programs will select the appropriate dark frame image from your "library" of dark frames and apply the correct one based on the exposure length and the temperature your imager is operating at.

Doing this part of the image reduction process at the time you are capturing your images offers several benefits. First and foremost, this is one step in the process that you will not need to accomplish during the image processing routines you will undertake after you have captured your raw data. Secondly, the imaging program will most likely take a series of dark frames at each exposure length and then perform the averaging routine automatically, so that it saves a ready-made dark frame for immediate use. If you are doing dark frame calibration manually, you will need to create a master dark frame from an average of multiple dark frames stacked together – more time spent during the image processing routine. Lastly, the images that immediately appear on your imaging display will not include these hot pixels and false data. This will make the images you see on your imaging display that much more visually appealing. Additionally, you will not mistakenly use a hot pixel as a reference point for frame stacking or worse yet, for auto-guiding. To avoid this situation, you will want to take dark frames for your auto-guiding imager as well as your camera that you are using for taking images. For your auto-guiding imager you won't have to take very many dark frames, as you will only be taking exposures of a few seconds at most.

If the imaging control program you are using doesn't offer immediate dark frame subtraction, or if you forgot to take dark frames before you started imaging and needed to take them afterwards, don't sweat it. Your image processing software will definitely have a calibration routine that includes dark frame subtraction along with light frame and bias frame calibration. When calibrating dark frames manually, you will need to have calibration frames prepared in advance. This entails selecting a number of dark frames at the appropriate exposure length and running them through the image processing routine in order to get one averaged and combined master dark frame. You will then select this master dark frame for use in the calibration process (Fig. 7.3). Calibration of your raw data images is always the first step you want to take in the image processing phase, prior to any other processing steps that you will perform on your images. Image processing will be covered in much greater detail in a later chapter.

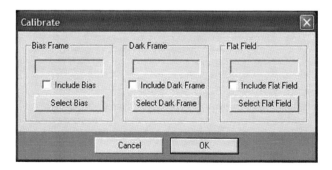

Fig. 7.3 Your image processing program will allow you to apply calibration frames as part of the processing procedure

Light Frames

The great thing about telescopes is their magnification power. The whole idea behind a telescope is that it takes objects that are so far away that they would be impossible to see with the naked eye and magnifies them enough to view through your telescope's eyepiece. With your CCD imager, this magnifying magic can be even greater. A 5-mile wide crater on the Moon (223,000 miles away) can be easily seen when imaged with a CCD camera (Fig. 7.4). The remains of a supernova that only takes up a fraction of an arc-minute of sky can be imaged so that it takes up a good portion of the frame in your imaging display (Fig. 7.5).

So what happens if the object you are imaging is as close as your telescope lens? The answer is the same; something that would be close to impossible to see with the naked eye is magnified to such a degree that it becomes very visible.

As you image distant objects in the night sky, light streams through your telescope's lens. The telescope not only magnifies the light coming through the lens, it also magnifies anything that may be blocking the light coming through your telescope lens, for instance dust, dirt, and fingerprints. Given the magnification power of today's telescopes, a fingerprint on your telescope's lens will show up as a giant gray or black object on your imaging display – and on your images, too (Fig. 7.6)! These flaws in your images will tend to vary in size, depending on the size of the

Fig. 7.4 A 5-mile wide crater on the Moon can be easily seen when imaged through a telescope

Fig. 7.5 The Crab Nebula (M1) only takes up 420 arc-seconds of sky as seen from Earth

Fig. 7.6 Through your imager, a dust mote appears to take up a huge swath of the sky

obstruction and also depending on which lens or mirror surface the obstruction is sitting on. Microscopic pieces of dust will show up as smaller visible flaws in your images.

Once again there is a way to fix these flaws through the calibration process. The fix for these kinds of flaws is by taking and applying what are called light frames or flat-field frames. The only way your image processing program can use these flat-field frames to fix or calibrate your images is if you take these frames with your imager in the exact same position and at the exact same focus and magnification that you have taken your images. For this reason you will want to take a set of light frames after each set of images you take, before you make any changes to your setup whatsoever.

This is an extremely important concept! If you change your focus at all, the flaws will still show up, but they will be a different size and will not show up on the same pixels that they do in your images. Likewise, if you make any changes to your setup, the flaws will again fall on different pixels, and the image processing program will not be able to negate the effect of these flaws in your images properly; trying to do so would in fact end up making your images look even worse!

So what exactly is a light frame or flat-field frame? As you can see in the image below (Fig. 7.7), a light frame has very even values across the image (a "flat" field), with the exception of any of the flaws, dust motes, fingerprints, etc., that you are trying to remove from your final image. These flaws show up as pixels that are darker than the rest of the pixels in your flat-field image.

When you calibrate your images using the light frame, the imaging program will automatically adjust the level of each pixel in your images based on the median

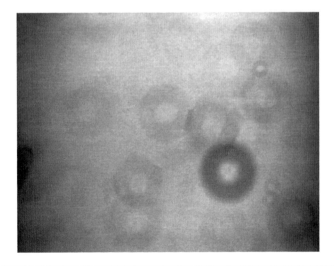

Fig. 7.7 A light frame has a flat, even field except where shadows from dust, dirt, finger-prints, or other obstructions appear

level of brightness in the overall light frame. If the median value of the pixels in your light frame is equal to 3,500 and the value in an individual pixel is only 2,500 because of a shadow created by a dust mote or other flaw, the image processing program will add 1,000 units of light value to this specific pixel in each of your images, thereby compensating for the false shadow. The end result is that each pixel's data value is adjusted to remove any of the flaws mentioned above, leaving only the data that you wanted to capture at the correct light values.

Here is another area in which one-shot color imaging saves you time. When you image through separate color filters, you need to take a separate set of flat-field images through each filter. Each filter can have dust, fingerprints, or other flaws on it, so a separate flat-field calibration image has to be applied in order to remove the specific flaws from the images taken through each filter. With a one-shot color image, a single set of flat-field images is all that is required and needs to be applied to only one set of images prior to separating the image into the individual color components for processing. This is one of the most important steps with a one-shot color camera – you calibrate the images before separating out the individual colors!

As will be discussed in a later chapter, during image processing, you will want to break the color image into its separate color components in order to adjust each color individually. Later you will reassemble the separate colors, perhaps along with a luminance image (also extracted from the same one-shot color images), back into a single RGB or LRGB image. When you break out the raw image into its separate color components, the computer program will use a series of interpolations that will also multiply any flaws in the image into the adjacent pixels. Given this fact, it is impossible to properly apply a flat-field image, or any other calibration image, after you have separated the individual color components from a one-shot color image. This fact makes it extremely important to apply the flat-field to each image in your series prior to separating them! This will ensure that only the correct data gets interpolated and will greatly enhance the quality of your final image.

There are several ways to go about capturing flat-field frames. One of the simplest ways is to point your telescope at a blank white wall, or a wall with a white sheet hanging on it. The wall needs to be well lit, however, so you need to have some very good lighting shining at the wall or the sheet. As the name implies, the lighting needs to be even across the area being imaged so you produce a "flat" field with no variations. The only variations you want to capture are the flaws that are inherent on your telescope lens or other surfaces that may be in your imaging path.

Getting this lighting even is sometimes tricky. Many amateur astronomers will actually wait until the following day and capture these images using a sunlit backdrop, again, making sure there are no shadows or other light variations in the field of view when capturing these frames. Unfortunately, since the light frames need to be captured using the exact same setup – magnification, Barlow lenses, focal reducers, focus position, etc. – taking these frames the following day precludes you from making any changes to your imaging setup throughout the night. This usually limits you to capturing images of a single object each night, or at best taking pictures of different objects but with the same setup.

Fig. 7.8 A light box enables you to take flat-field images immediately following the imaging of a specific target so you can make configuration changes when you move onto the next target

Of course with a little up-front planning you may be able to select targets that are of the same general size that you would want to image with the same setup anyway. That is an excellent method of saving time. Usually, though, you will want to capture the light frames as soon as you are done imaging your target. Then you are able to move onto another subject and change your system configuration any way you choose. Just be sure to take more flat-field images in that configuration when you are finished imaging the next target.

Another way to capture flat-field images is by draping a white sheet or a white T-shirt over the end of your telescope, but probably the best way to accomplish taking flat-field frames is through the use of a "light box". A light box (Fig. 7.8) is a portable flat-field generator that can be attached to the end of your telescope in order to take light frame images.

Incorporating a white piece of Plexiglas at one end of the box, small lights are spread out around the inside of the box to provide an even field of illumination on the Plexiglas. You will then take images of the Plexiglas in order to capture your flat-field frames. Often running on battery power, a light box has the advantage of being able to be used in remote imaging locations when a blank white wall or a convenient place to hang a sheet is not available.

In most instances, using the light box to capture your flat-field frames also allows you to do so without having to reposition the telescope, thereby avoiding any chance of the primary mirror shift that could ruin your light frames for use in the calibration procedure. A light box lets you capture your flat-field images quickly and easily and then move on to another target so that you can image as many different targets as you have time for in a single evening.

A simple light box can be manufactured at home with some foam-core board, a hot glue gun, some Plexiglas, and a few small light bulbs and wiring, all available at your local hardware store. There are plans available in several locations on the Internet, or you can find plans, kits, or assembled light boxes at www.digitalspaceimages.com. If you plan to use the light box out in the field, you may want to consider making your box out of a more sturdy material, duct-taping the edges, or at least weatherizing the outside of the box with some paint or acrylic sealer spray to avoid damage from dew or from simple wear and tear during transport and use.

Just like when taking dark frame images for image reduction purposes, you will want to take a series of light frames (typically a minimum of five images) and then use an average-combine method to create a master light frame for the calibration of your images. Unlike taking dark frames, light frame images do not need to be taken at the same exposure length of the images you will be calibrating them with. Instead, you simply need to take them so that the bulk of the light captured in your flat-field image is somewhere between 35% and 65% of your imager's dynamic range. This will be explained in more detail in the upcoming chapter on histograms. The best part is that it will take far less time to capture five flat-field images than it takes to capture several sets of dark frame images.

Although the math involved in using a light frame to calibrate your images isn't exactly straightforward, the good news is you won't have to worry about that. When you have your master light-frame created, your image processing program will have a way to easily use this master light-frame in your calibration routine (Fig. 7.3). Once you have applied your master dark-frames and your master light frames to your images through the calibration process, you will be ready to stack several of your images together to increase the signal-to-noise ratio. You will separate your images into their individual color components, process these individual component images, and then combine them into one-color master images. After you have combined your individual one-color master images into a full color image, additional processing can be performed to enhance your image anyway you want in order to produce your final image. Much more information will be provided on image processing in a later chapter.

Bias Frames

The last type of calibration image that can be used for image reduction purposes is known as a bias frame. A bias frame is basically an image taken at the shortest possible exposure time (a mere fraction of a second) that shows any noise inherent in your imager that is not related to dark current. Bias frames only need to be used if you have not taken dark frames at the same exposure length that your images are taken. As stated earlier, it is highly recommended that you do, in fact, take your dark frames at the same exposure lengths as your images. Pixels across your imaging chip can react quite differently at various exposure lengths and the corresponding temperature differences that will occur.

Nevertheless, in order to save time taking a series of very long dark frame exposures, some experienced astro-imagers will take a shorter dark frame and then "scale up" the results to match the exposure length of the images they are calibrating. For instance if you are looking at a 30-min exposure, you could potentially take 10-min dark frames and multiply the value of each pixel by three times in order to get a 30-min scaled dark frame. Again, this approach is not recommended, as it is only an approximation of the actual dark current values involved with the longer exposure length, but it can be done and will provide better results than not performing any dark frame image reduction at all. The problem with multiplying the dark frame pixel values by three times in the example above is the existence of noise in the imaging chip that is not related to exposure length.

Just like when taking dark frames, bias frames are taken with the "lens" cap on or with the end of the telescope covered so that no outside light can enter the CCD imager. If you take a bias frame at an exposure length of one-tenth of a second, your average pixel value may end up at a value of 30, even though not enough time has passed to build up any dark current values in your pixels. This is the noise inherent in your imaging chip, and it exists at the same level regardless of how long the exposures you are taking happen to be. If you then take a dark frame of 10 min in length, your average pixel value may come in at 530 units. The 500 units of this value are due to dark current at the 10-min exposure length, and the remaining 30 units are a result of the bias. When you scale up this dark frame for use in a 30-min image, you will multiply the 530 units by 3, resulting in a scaled dark frame average pixel value of 1,590.

If you turn around and take a 30-min dark frame for comparison, you will find that the average pixel value will be closer to 1,530, The 1,500 units of this value are from the thermal noise from dark current readings at this exposure length, and the remaining 30 units are the bias that does not change with exposure length. As you can see by simply multiplying the value of the 10-min dark frame, you have inadvertently multiplied the value of the bias which, again, does not increase with exposure length. The way around this problem is to subtract the bias level from the 10-min dark frame before you scale it up by 300%. This is how and why bias frames are used in image reduction.

If you are going to be scaling dark frames to calibrate longer exposure images you will need to use bias frames in order to get the best results. Image processing programs will have an easy way to select bias frames for use in image reduction during your calibration routine (Fig. 7.3). Again, several bias frames should be taken and median-combined to create a master bias frame for this purpose. Once again, however, it can't be overstated – the way to get the best possible results in your final images is to simply get in the habit of taking dark frames at the same exposure length of your images.

Now you know what to do before you start imaging – set up your equipment, take dark frames, and get your telescope focused for use with your CCD camera. You also know what to do immediately after your imaging session – take flat-field frames and, at your leisure, calibrate your images and process them into the final version depending on your goals and tastes. Now it's time to look at what's in between: the imaging process itself.

Chapter 8

Taking Exposures and Auto-Guiding

Once you have your 'scope in focus, your target image framed properly, and your dark frames taken with your imaging program set to automatically subtract the dark frames for you, you are ready to start your imaging session. The next decision you will have to make is how long of an exposure to use for your images.

Exposure Times

The proper exposure time varies based on a number of different factors. The first consideration is the brightness of the object you are imaging. Bright objects such as the Moon, or the larger planets (Jupiter and Saturn), require very short images. Because they are reflecting lots of sunlight that is focused directly onto your imaging chip, the photons add up very quickly and can easily saturate your image. By taking very short exposures, usually for only a fraction of a second, you will capture excellent data and the end result will be high-quality, beautiful images. For more faint objects (as seen from Earth), such as galaxies, nebulae, and star clusters, longer exposures will be required in order to capture enough data to make your images come to life and show as much detail as possible.

When it comes to these faint deep-sky objects, the general rule of thumb for exposure time is the longer the better. There are a few factors that can limit the length of exposure time. The biggest factor in long exposures is light pollution. If you live in a rural area, faraway from the lights of a city, the length of time you can expose your images for is limited only by the steadiness of your mount, the precision of your polar alignment, and the quality of your auto-guider (if you are

L.A. Kennedy, *One-Shot Color Astronomical Imaging*, Patrick Moore's
Practical Astronomy Series, DOI 10.1007/978-1-4614-3247-0_8,
© Springer Science+Business Media New York 2012

using one). However, if you live in or anywhere near a city, light pollution can quickly saturate your image in only a few minutes or less. The good news is that there are several ways you can get around this problem.

The easiest way to combat light pollution in your images is to take lots of shorter exposures and "add" them or "stack" them together later. Although a longer exposure is always better, because of exponential factoring of the desired signal and its inherent ratio to "noise" in the stacking process, stacking several exposures together can yield almost as good of an image as capturing a longer exposure will. Of course several longer exposures would ultimately be better, but multiple short exposures can produce very nice images as well.

Another way to circumvent the problem caused by light pollution is to use a specialized light pollution filter. This is a filter that fits onto your imager that only lets specific wavelengths of light through.

Objects in the night sky typically put out light in specific bands of the light spectrum, while artificial objects, such as streetlights, put out their light in different bands of the light spectrum. By blocking all bands of wavelengths from areas outside of stellar light wavelengths, a light pollution suppression (LPS) filter can eliminate the "wrong" spectrums of light from entering your imager. Although this also tends to cut off some desirable spectrums of light, particularly in the infrared band, the benefits of using this kind of filter can far outweigh the downside that this loss of bandwidth causes.

If light pollution is not a factor for you because you are lucky enough to live in an area that is not affected by light pollution, or if you are using a light pollution suppression filter, then the next limiting factor for your exposure time is the saturation of your imaging chip. Often your imaging subject will be near stars that are much brighter than your target itself. While trying to gather enough data of your imaging subject, the brighter stars in your image can saturate the particular pixels that cover that area of the image.

When these brighter stars are outside of your target image, this is not such a big problem, and there are ways to remove these saturated areas from your final image that will be covered in a later chapter. When the bright stars are located inside your imaging subject, or if they happen to line up between your target and your imager, the saturation caused by these brighter stars become a little more problematic.

There are ways to deal with these situations, too, and these will also be covered in a later chapter, but in order to reduce the amount of processing required after the images are downloaded, it is best to minimize the amount of saturation caused by these brighter objects. The most effective way to do this is by limiting your exposure time to keep these objects from saturating the chip and compromising the data of your imaging subject.

Probably the best way to ensure you are not saturating the pixels in your images is to take a test image. You can still use this test image if it looks OK; just let the imager keep running, but check how the first image looks after it reads out of the chip and downloads to your imaging display on your computer screen. If it appears that pixels have in fact been saturated on your first image, then you may want to restart your imager with a shorter exposure time, again checking the first image that downloads for saturation and picture quality.

You can also look at the image's histogram – a graph showing the counts of the number of pixels that have collected the same number of photons. Histograms and their use will be covered in depth in the following chapter.

Before you start taking your first image you will want to select a star or two as a reference point for your imaging program to use in order to stack your images together. Whether you are taking short exposures or long exposures, you will want to take a few of them in order to combine them together, or stack them. Choosing these reference points gives your imaging-control program a means to make sure these images are stacked so that they line up in exactly the same way, thereby compensating for any minute movements of the subject in your images.

These reference points will only be used during imaging for the display of the images that you are currently taking. When you get to the image processing phase, you will choose reference points again in order to combine several images into one version.

There are, of course, several reasons why you will want to take multiple images. Some images will turn out better than others due to different seeing conditions during the times you are taking each specific image; or cosmic rays could potentially strike your imaging chip, leaving an ugly spot in your image; and even more likely, meteors, planes, or even passing satellites can cross through your CCD camera's field of view, resulting in long streaks of light that will be difficult to repair (Fig. 8.1). Having several images to choose from will let you scrap any that might be ruined, or may be of lesser quality, and yet still have enough images left over to work with during the image processing.

Fig. 8.1 You will want to take multiple images in case one or more of your images is ruined by a passing plane or satellite

The other main reason to take several images regardless of the length of exposures you are taking is to increase the signal-to-noise ratio when the images are combined during the image processing procedure.

Signal Versus Noise

So what is the signal-to-noise ratio, and why is it so important? As mentioned before, the "signal" is the data in your image that you are trying to capture. The "noise" is unwanted "false" data, or uncertainty in the brightness levels of the pixels, that you would rather not capture. The signal-to-noise ratio is the way to measure the amount of noise in your images. Although in reality it's a little more complicated, for purposes of this discussion you can think of the signal-to-noise ratio in the following terms:

$$\frac{Avg.\ Signal\ Pixel\ Value}{Avg.\ Noise\ Pixel\ Value} = Signal - to - Noise\ Ratio$$

So if the average value of the pixels with your intended data is 2,000 and the average value of the pixels with the unintended noise is 100, then your signal-to-noise ratio (abbreviated as S/N) is equal to 20. In the end it doesn't matter what the exact mathematical equation for signal-to-noise is. All that matters is that you understand that signal is good and noise is bad. In order to improve image quality you will want to maximize the signal in your images and minimize the noise. So how do you accomplish these tasks?

There are really only a few things you can do to accomplish both tasks. In order to maximize the signal in your images you want to make your exposures as long as you possibly can, subject to the quality and steadiness of your equipment, the accuracy of your polar alignment, and the saturation of your imaging chip. You will also want to make sure you are using the image reduction techniques previously discussed to calibrate your images. Lastly, as mentioned, you will want to take multiple exposures and combine them, which will increase the larger signal values faster than it increases the smaller noise values (more on this technique later).

In order to minimize noise, you will want to do your imaging at the lowest possible temperatures. Any cooling you can do of your CCD imager, or any imaging you can do outside where temperatures are cooler, will provide exponential noise reduction in your images. For most CCD cameras today, the cooling factor is what it is, but often there are additional accessories that you can use, such as heat-sink fins or an attachable fan that will help to lower the operating temperature even further. Some CCD cameras have an adjustable setting for their cooling operations. If your imager has this adjustable setting, keep it set to cool the imager as much as possible without setting it at 100%. (You will want a little wiggle room to deal with potential temperature spikes if need be).

One way to improve signal and minimize noise at the same time is to do your imaging under the darkest skies possible. This allows you take longer images and collect more data (signal) in these images. It also reduces noise by avoiding light pollution and chip saturation, both of which result in added noise. Although this isn't always convenient or even possible for everyone, anything you can do to minimize the effects of light-polluted skies will help your images drastically.

Saving Your Images

The other decision you will need to make is how to save your images. Most imaging programs will offer a variety of ways that you can save your images along with different formats that you can save them in (Fig. 8.2). The best format to save astronomical images in is called FITS (Flexible Image Transport System). This format was specifically designed for use in scientific imaging. It offers several benefits for use in astronomical imaging that other file formats do not, including header information regarding the image itself.

FITS is commonly recognized by many astronomical image-processing programs that you will be using to manipulate your images into their final form. There are also conversion programs that can be downloaded off the Internet for free that will allow other computer programs that do not recognize FITS files to be able to utilize the

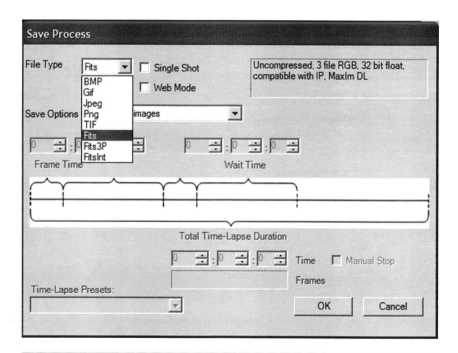

Fig. 8.2 Your imaging control program will offer a variety of formats you can save your images in

.fits file format. If you need a program to view and manipulate FITS files, you can download the free FITS Liberator from NASA and the European Space Agency at the following website: http://www.spacetelescope.org/projects/fits_liberator.

When determining which method to use in order to save your images onto your computer, you will always want to choose to save the "raw" images. When using any type of CCD camera, you will not want to save only the "composite," or combined output, images. All it takes is one bad image being used in the composite to ruin your entire imaging session. A little wind, a passing cloud or an airplane, even a meteor streaking through your image could all affect a single image or two, which in turn will affect the composite image. You will want to have the option of examining each image before it is used to create the final version of your image stack. Otherwise these flawed images get combined with your good images and the flawed data is averaged, median combined, or otherwise incorporated into your final image, making it all but impossible to fix after the fact.

With a one-shot color camera, it is especially critical that you are saving the raw images (Fig. 8.3) as opposed to the images in which the imaging program automatically separates the individual color components and only saves these images. As mentioned in a previous chapter, due to the Bayer array and the way that one-shot color imagers take red, blue, and green images all at the same time, the imaging control program needs to interpolate data for each color based on the data in the

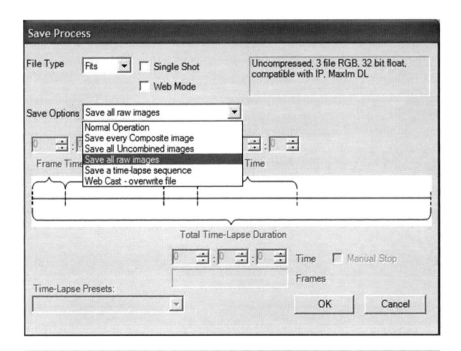

Fig. 8.3 Your imaging control program will offer a variety of methods for saving your images

pixels that image in one specific color. It is critical, in order to get the best possible image quality, that you calibrate the original raw images before the images are separated into the individual color components. Again, this point cannot be stressed enough! This is one of the key things to remember in order to image effectively using a one-shot color camera. Most books or articles on image processing miss this point and merely discuss how to process color images taken through individual color filters with a black and white imager. They will state that each colored image component needs to be calibrated separately (which is correct for these types of black and white filtered images), but with a one-shot color image, the image reduction process *must* occur prior to separating the color components or additional flaws will creep into each colored image that cannot be removed correctly. This will result in more noise and a lower signal-to-noise ratio!

Auto-Guiding

As mentioned previously, one way to improve the signal-to-noise ratio and improve your image quality is to take longer exposures. In order to take longer exposures you will need two things. One is accurate polar alignment of your telescope, which was covered in a previous chapter. The other is guiding, which will be discussed now.

Guiding is the process by which small adjustments are made to keep your telescope pointed to the exact same spot in the sky, thereby keeping your targeted subject at precisely the same spot on your imaging chip. This can be done manually, and in fact this is the only way that it used to be able to be done.

To do manual guiding for your imaging, you will need an off-axis guider. This is a piece of equipment that slides into your telescope's eyepiece holder, or into the main opening of your telescope's central axis. The off-axis guider, in turn, has two eyepiece holders, one on the opposite end of the guider where you attach your CCD camera so that the light path of the telescope goes straight through onto the imaging chip. The other eyepiece holder is designed at a 90° angle to the axis of the guider ("off-axis") and utilizes a mirror set at a 45° angle to "pick-off" a small portion of the stream of light and send it through the off-axis eyepiece.

A lighted-reticule eyepiece is set into the off-axis eyepiece holder. You will also want to use a Barlow lens in front of the lighted-reticule eyepiece to increase the magnification and thereby the sensitivity and accuracy of the manual guiding process. Using this setup, a guide star is centered in the crosshairs of the lighted-reticule eyepiece, and while imaging the observer keeps a constant eye on the guide star, making manual adjustments to the pointing position of the telescope in order to keep the guide star centered in the crosshairs of the eyepiece.

Manual guiding requires a lot of time spent hunched over the eyepiece monotonously watching the guide star while making manual adjustments to the telescope when you see the guide star drifting out of position. You must watch the guide star through the eyepiece for the entire length of your imaging exposure, and if you are taking multiple images (which you should do), then you must keep watching the guide star and make the necessary corrections for the entire length of your imaging session.

Now, with the advent of computerized telescopes and relatively inexpensive CCD imagers, there is a better way to accomplish guiding; this is the process known as auto-guiding. In order to perform auto-guiding, your computerized telescope must have the functionality that allows it to be controlled by a CCD imager through the imaging-control program. Refer to your telescope's instruction manual to confirm that your telescope does indeed have this functionality built into it. Most modern computerized telescopes do. If you are not using a computerized telescope, then you may still be stuck doing manual guiding the old-fashioned way.

The concept in auto-guiding is to use a separate CCD camera, or a dedicated auto-guider, find a guide star on the imaging chip for the auto-guider to track, then let the camera or auto-guider control the movement of your telescope in order to keep the guide star positioned on the same spot on the guider's imaging chip. By keeping the guide star in the same position you will also be keeping the subject you are imaging in the same position on the CCD camera you are imaging with.

In most cases auto-guiding will require not only a second CCD camera, or a dedicated auto-guider, but also a second smaller telescope. There are some higher-end model CCD cameras that have auto-guiding capabilities built right into them, but these types of dual-function self-guiding imagers have their own issues to deal with. Most of these will use a prism to pick off some of the light coming in through the telescope so they can only "see" a small portion of the total field of view for guiding. This creates a problem when trying to find a suitable guide star to use for tracking. Other self-guiding CCD cameras will use half of the captured data to do the imaging and hence require twice the exposure length in order to get the same amount of data as a regular CCD camera. If you don't own (or want to purchase) a second telescope, you might consider using the off-axis guiding technique that performs the same function as it does with manual guiding. You can put a second CCD camera, or a dedicated auto-guider, into the off-axis guider's eyepiece holder and use this arrangement to do the auto-guiding for you.

Most likely, though, you will want to use a second telescope, known as a guide 'scope, attached to your main telescope, along with a second CCD imager or a dedicated auto-guider, set up on the guide 'scope. This configuration is what you will then use to auto-guide your imaging telescope.

This approach has many advantages over the prism-based or self-guiding methods and will be well worth the additional money you will spend on this setup. First, you will be much more likely to be able to find an appropriate star to use for auto-guiding. If you are just picking off a small portion of the field of view with a prism or a mirror, then you have to have a little luck, or, more likely, you will need to adjust your field of view so that an appropriate guide star is present in the portion of the light path coming through the prism. With a second CCD imager or auto-guider using its own full field of view, having a suitable guide star somewhere on the imaging chip is much easier to accomplish.

In order to do the guiding, your CCD imager will need to use a reasonably bright star as its reference point. The star needs to be bright enough for you to be able to see it with fairly short exposure intervals. The brighter the star, the shorter the exposures you can take and still have your guider "see" the star well enough to use it as a reference point for the guiding corrections. Many times these bright stars aren't conveniently located in the areas surrounding your imaging subject. It isn't fun to have to make a choice between properly framing your subject on your imaging chip and having to sacrifice the ideal framing in order to land a guide star on your auto-guider's chip. With your auto-guiding imager seeing an entire field of view, you are much more likely to be able to find a good star to use no matter how you have your target framed on your imaging chip.

The separate guide 'scope and auto-guider setup will also provide an easier way for your auto-guider to be able to control your telescope. Using this method you can align your guide 'scope parallel to the axis of movement of your imaging telescope so that when your auto-guider tries to move your telescope based on what it perceives as the correct direction, your telescope moves in that exact direction. If the alignment of your guide 'scope and your imaging telescope's axis is slightly skewed, the auto-guider may have trouble making the correct adjustments in the proper direction to bring the centroid, or center, of the guide star back onto the same pixel where it started.

Some auto-guiding control programs don't have any problem making the appropriate corrections no matter how the guiding imager is set up, but many require the auto-guiding imager to be completely squared up with the right ascension and declination axes of the imaging telescope. This squaring up is known as orthogonality, and you can check your imaging/auto-guiding control program's instruction manual to see if it requires an orthogonal arrangement or not.

For your auto-guiding camera, you can use just about any model of CCD imager, but you will want to use one that has enough dynamic range to see the same field of view as the imaging CCD; for instance you won't want to use one of the more basic imagers that are designed strictly for use in lunar and planetary imaging. On the other hand, you don't need a high-end imager for auto-guiding purposes either. The auto-guider does not have to be a one-shot color imager; black and white will do just fine. There are also CCD cameras known as dedicated auto-guiders that are made specifically for use in auto-guiding. By not having all of the bells and whistles needed for actual imaging built into them, they can be a bit more affordable when all you are going to be using it for is auto-guiding.

For the second telescope, you obviously need something smaller than your main imaging telescope, since this second telescope will need to be "piggybacked" onto your larger telescope (Fig. 8.4). Although it helps the accuracy of your auto-guiding if the second telescope is approximately the same focal length as your imaging telescope, this is not absolutely critical. It's best to use a guide 'scope that is no less than half of the focal length of your imaging telescope, but you can ultimately use any size telescope with varying degrees of accuracy.

In fact, you may want to use a guide 'scope that is completely different from your main imaging telescope, thereby offering you a wide array of imaging options with your combination of equipment. Having one telescope with a slow focal ratio for high-magnification imaging and another telescope with a fast focal ratio for wide-field

Fig. 8.4 A smaller telescope can be piggybacked onto your larger telescope along with another CCD camera for use in auto-guiding

imaging is probably the ideal situation. You can then alternate between the two different telescopes, using one telescope for imaging and the other for auto-guiding, depending on what type of target you choose to image on a particular evening.

Whatever type of telescope you use, it will need to be firmly attached to your main telescope to avoid flexure, or movement of one telescope in relation to the other, which will cause unnecessary adjustments to be made by your auto-guiding program. You will also want to make sure that your telescope/auto-guider is pointed to the same area of the sky as the CCD camera you are using for imaging. If not, you can still end up with field rotation in your images that will be very difficult to compensate for after the fact.

The auto-guiding process is relatively straightforward. You will attach your auto-guiding imager to the telescope you are using for auto-guiding purposes. You will need to connect the auto-guider to your computer and in turn connect the computer to your imaging telescope, although some auto-guiders can attach directly to the mount itself, depending on the type of equipment you have. You will need to focus your guide 'scope with the imager or auto-guider so that it will have a clean, crisp star to work with. You will then go into the auto-guiding routine in your imaging-control program and select a bright star to guide with. The imaging-control program will then calculate the center, or centroid, of the guide star. After this, the imaging-control program will make some test movements to see which direction, and how long the correction commands need to be, in order to move the telescope to the proper location.

Fig. 8.5 An auto-guider requires several settings and other pieces of information to be assigned for use in auto-guiding

Your imager may or may not be set up to move in the same direction as you see on the imaging chip. Keep in mind that some telescopes provide an inverted image depending on the number of mirrors, if any, that are incorporated into the telescope's design. If the field of view is inverted, a command to move the telescope to the right may result in the star moving to the left on the auto-guider's imaging chip. Similarly the direction of required movement will reverse depending on which side of the meridian you are imaging on. The auto-guiding test routine will recognize this and will send the necessary correction commands to your telescope in order to move the guide star in the right direction.

A basic auto-guiding control program will require you to provide the focal length and aperture measurements of your guide 'scope. There are a couple of other settings you will need to make in order to get your auto-guider operating in the most efficient manner (Fig. 8.5). First you will need to decide on the exposure length to which to set your auto-guider. Your auto-guider will need a bright star to work with, and the brighter the star is, the shorter the exposure lengths you can use.

Since your imager has the ability to take images that are a mere fraction of a second in length, if you have a bright enough star on your guiding chip, it may be tempting to use these very short exposure intervals to calculate your auto-guiding movements. This is typically not the way you want to go, however. The star's appearance and location on your auto-guider chip can often be a bit deceptive, due to atmospheric turbulence and less-than-optimal seeing conditions. Although these movements will also show up in your images, they do not necessarily make imaging impossible; they only make the stars in your image spread out over multiple pixels. If your auto-guider makes an unnecessary correction for each of these movements, however, the stars will be spread out even more on your images and could end up being oblong instead of circular.

The best way to deal with this problem is to use longer length auto-guider exposures, somewhere between 3 and 7 s, or even longer if needed. If you have a good, steady mount and your polar alignment is fairly accurate, this should be more than enough time to catch any needed corrections in your telescope's movement, and you will avoid over-correcting based on the less-than-ideal seeing conditions.

The other setting you will need to decide upon is known as the "correction gain." The correction gain is a setting that tells your imaging control program how big of a percentage of the needed correction to make at one time. Intuitively you may be wondering why you do not make a full correction each and every time it is needed. The reason is that some mounts are more reactive to correctional adjustments than others, depending on a number of factors – how well the mount and equipment is balanced, how much play there is in the telescope's drive gear, etc. Many mounts also react differently depending on what area of the sky they are pointed at, moving a greater distance when imaging near the zenith than they do when imaging near the horizon.

In practice, making a full correction each time one is needed can often lead to over-correction and creates the need for a compensating correction in the opposite direction. This constant back and forth movement will spread your stars out on your images even further than they otherwise would be. In order to get a handle on this setting, you will want to experiment a bit starting with a correction gain setting of somewhere between 50% and 75%. With this setting, as your auto-guider detects the need for a correction, your imaging control program will move the telescope one-half to three-fourths of the way needed to put the centroid of the guide star back to its original position. When the auto-guider downloads the next image it will recalculate the amount of correction needed and again move the telescope one-half to three-fourths of the amount needed to reposition the guide star. Experimenting with the correction gain setting will enable you to determine the best setting to use given the functionality of your mount, the specific seeing conditions you are imaging in, and all of the other factors involved in creating the need for a correction in the first place.

Once your auto-guider has performed its test movements and begins the auto-guiding process, you should notice after a short period of time that your auto-guider will probably "settle in" and will need to make fewer and smaller corrections. This is when you will want to begin capturing your images with your imaging CCD; sometimes this can take several minutes depending on the quality of your mount. After you are done capturing your images, be sure that you turn your auto-guider off from controlling your telescope before you go to slew your telescope to another location.

When you are ready to image another target, select a new guide star for your auto-guider and turn the controls on, and let the auto-guiding program run through the test movement process again. You will then be ready to capture your next set of images of your next imaging subject.

Fig. 8.6 An auto-guider will use the centroid of a guide star to calculate the corrections required to keep the guide star centered on the original pixel it started on

Ideally, if you have all of your settings correctly applied, have good polar alignment, and are using a decent mount, you auto-guider will keep the centroid (Fig. 8.6) of the guide star within ±1 pixel in movement, ±0.5 pixels would be even better. The faster your focal ratio is, the easier your auto-guider will be able to accomplish this task as each pixel will cover a greater area of the sky (remember the arc-seconds per pixel measurement). Longer focal lengths and the resulting slower focal ratios require much more precision from your auto-guider.

If your telescope is equipped with it, another way to make it easier for your auto-guider to accomplish good guiding is to turn on your Period Error Correction (PEC) setting. Using the PEC setting will allow the telescope to make its own positional adjustments to offset any unwanted movements caused by slight errors in your telescope's worm gear. This gear is what the telescope uses to track objects across the night sky, and small imperfections can cause your telescope to shift based on slight inaccuracies in the gears. The Periodic Error Correction compensates automatically for these imperfections so your auto-guider doesn't have to.

You can also make adjustments to your backlash setting, which speeds up the movement of the telescope for a fraction of a second when you are changing from one direction to the other while auto-guiding. There is often some play in the motorized gears that arises when switching the direction of the correctional movement, and the backlash adjustment is designed to take up the slack "instantaneously." This allows your auto-guider to make the appropriate positional adjustments much more quickly, before the guide star has a chance to move too far out of position.

Chapter 9

Histogram Display and Stacking

Histograms

There are many different tools available and lots of information at your disposal that will aid you in figuring out how to improve your imaging and image quality. Probably the most useful tool and source of information that you can use is known as the histogram display. A histogram is a graphic representation showing a visual picture of the distribution of data. For imaging purposes, the histogram represents the distribution of the photons, or light, being captured by all of the pixels on your imaging chip. This is a very powerful tool to help you make sure you are capturing your data in the best possible area of the dynamic range of your imaging chip.

As stated, a histogram is a visual graph of the amount of data (photons) present in each pixel across the image as a whole (Fig. 9.1). The horizontal axis of the graph contains value counts (# of photons present in each set of pixels) ranging from zero to the maximum capacity of your pixels. The vertical axis is the number of pixels that contain the same value count. On the far left side of the histogram graph is a vertical bar showing the number of pixels with a value of zero. Just to the right of that bar is another vertical bar showing the number of pixels with a value count of one, the next bar counts the number of pixels with a value count of two, and so on and so forth.

Darker areas of the image ("empty space") have very few photons collected in their pixels, and the counts of these pixels show up on the left side of the histogram. As the number of photons collected per pixel increases, you can see an obvious spike as you move to the right across the histogram graph. These are the pixels that contain the data for dark, empty sky. Theoretically, these pixels should have a value

L.A. Kennedy, *One-Shot Color Astronomical Imaging*, Patrick Moore's
Practical Astronomy Series, DOI 10.1007/978-1-4614-3247-0_9,
© Springer Science+Business Media New York 2012

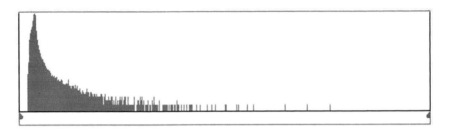

Fig. 9.1 A histogram is a graph showing the number of pixels with the same value at each value count across the whole range of the imaging chip's capacity

of zero, but this is rarely the case due to the effects of skyglow in your images. This light pollution causes even the darkest sky areas to read some sort of brightness value, but as will be shown below there is an easy way to compensate for this problem. You will see an obvious drop off on the right side of this spike as the count of the pixels that are brighter becomes less and less; this is usually the data representing your imaging subject, the data you are trying show in your images. Eventually the pixel counts will taper off to a very minimal count of the brightest pixels on the image (the pixels showing the centers of stars, etc.).

The histogram display is used in two specific processes involved with astroimaging. First it is used during the imaging session in order to adjust the display on your computer screen. The histogram display is used again during image processing to adjust the values that are eventually displayed on your final images.

During image capture, depending on the subject you are trying to image, you may need to adjust the display on your computer screen so you can actually see the target. Keep in mind a typical computer screen can only show 256 different color variations, while a CCD camera can handle 65,000 different variations or even more, depending on the imaging chip in your CCD camera. In order to display the full range of values on your computer screen, the monitor has to allocate 65,000 different values into the 256 variations it can show. Many times up to 90% of the values being allocated to the limited graphics of the computer display are variations of what should be pitch-black sky or the variations in brightness between different-sized stars. This doesn't leave a whole lot of the computer's graphic display to be allocated to the data showing the imaging subject. The solution to this problem is to use the histogram display to shorten the range of values being displayed on your computer screen.

Bright stars, globular clusters, and planets can be easily seen with a large range setting on the histogram display. If you are imaging nebulae or faint galaxies, on the other hand, you will probably need to display a much shorter range of data on your computer screen. Your imaging-control program will most likely have some sort of automatic contrast adjustment that you can select. Unfortunately, this doesn't always give you the best view of your targeted subject. Many celestial targets are very faint and only take up a small portion of the total data captured. By adjusting which portion of the data is displayed on the computer screen, you will be able to make these targets much more perceptible as you take your images (Figs. 9.2 and 9.3).

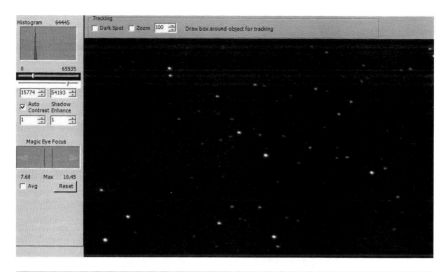

Fig. 9.2 Using the automatic contrast setting, the Veil Nebula is too faint to be seen on the computer's image display

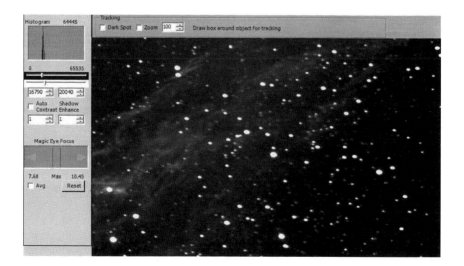

Fig. 9.3 When you shorten the range of data to be displayed manually, the Veil Nebula shows up very clearly on the image display

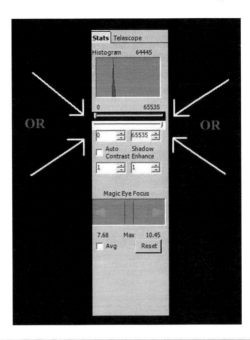

Fig. 9.4 The histogram display has two settings, either of which will control the range of pixels to be displayed on your computer screen

To do this you simply need to adjust the minimum and maximum display values, known as the black point and the white point. The computer program that you use to capture data and to display your images will give you the ability to set the minimum and maximum display value on the histogram graph, either by typing the data values into the program, or by moving the slider for the black point and/or white point (Fig. 9.4).

Setting the minimum display value, or the black point, where the histogram indicates, will cause the bulk of your data to start showing up on the left side of the histogram and make all pixels with a photon count below that number be displayed as dark or black areas of the image. Remember, most astronomical images are empty sky and should show up as dark pixels. You can see this as a sharp spike in the pixel counts as you move to the right from zero on the x-axis. Set the black point just to the left of this spike in data so that you display all of the skyglow pixels as black (Fig. 9.5).

On the other end of the graph, you want to adjust the maximum display value, or the white point, so that it allows the majority of your target data to be displayed. Wherever you set this maximum value at, any pixels with a photon count higher than this number will show up on your image as a pure white pixel. Move the white point to the end of the data curve where your imaging subject shows up. You should see an obvious drop off in the pixel count where only the centers of stars and the data for other very bright areas was captured (Fig. 9.6).

Fig. 9.5 The *black* point slider determines which pixels will be shown as black on the image display

Fig. 9.6 The *white* point determines the pixels that will show up as completely white on the image display (globular cluster M56)

Any pixels whose photon counts show up in between the minimum and maximum settings will be spread out across the range of values that can be displayed on your computer screen. It will take a little practice to get the hang of making these adjustments. While you are imaging, you can set the minimum and maximum wherever you desire, and it will not affect the data that your imager will capture. The data is not lost, so you can try the settings at various levels in order to display the image any way you want on your computer screen during your exposures. When you are processing your final images, it becomes much more important to find the right settings for the minimum and maximum of the histogram display in order to optimize the quality of your images. After you save the image, the data outside of the black and white points will be lost, leaving only the range of data that you want to appear in your image, so you need to be very careful where you set the black and white points during image processing. If save the new image under a different name, however, then you can always revert back to your original image if you want to make changes later on.

During the imaging session, you ideally want the bulk of the data from the imaging subject to line up in the first two-thirds of your histogram (Fig. 9.7). This allows enough room on either side of the desired data to set the black and white points appropriately. If your subject data does not fall in this range, you may want to consider taking longer exposures, assuming you don't already have any saturated pixels showing up in your image. If your subject data is showing up more than two-thirds of the way into your histogram, depending on what your imaging subject is, you may want to consider using shorter exposures.

Fig. 9.7 Ideally, the data you want to capture should land somewhere between one-third and two-thirds of the dynamic range of your CCD camera (The Triangulum Galaxy, M33)

The histogram display will be used both during image capture and again during your image processing routine. During the image processing phase, it is important that you select the appropriate black and white points for your image, as the data outside of these points will be lost. When setting the black point, you will want to set it far enough to the left of the data spike so that your empty sky areas show up as a very dark gray, as opposed to pure black. This is much more visually appealing and won't affect the range of available display levels by that much. Experiment with the white point setting to enable as much detail to be shown on the image as is possible. Keep in mind that this is a subjective exercise; there is no right or wrong setting. The only factor is how you think your images look their best, depending on what details you want to show.

Familiarize yourself with the histogram display, how it works, and what information it is showing you. This will be one of the most important tools in your arsenal to fine tune your images. As will be discussed later in the book, manipulation of the histogram will allow you to maximize the range of variations displayed on your images by applying a curve to the histogram display. Again, the histogram is a very powerful tool and an important source of information you can use to enhance your images.

Stack 'Em Up!

One of the biggest advantages of CCD imaging over traditional film imaging for astronomical use is the ability to take multiple images and then combine them during image processing. This feature allows you to collect data for as long as you want and then combine it all, multiplying the signal while reducing the amount of noise present in individual images alone. Stacking multiple images can also compensate for not being able to take very long exposures due to light-pollution, less than ideal equipment, or pixel saturation. Although longer exposures are always better, multiple exposures can be equally effective at capturing high levels of detail in your images.

When you are taking multiple exposures during your imaging session you should save each raw image separately as opposed to saving one combined image with all of your exposures included. You will want to do this for two very important reasons. First off, you will want to calibrate each of your raw images separately. By removing as much noise as possible in each of the raw images before combining them you will multiply the signal you want to keep without multiplying the noise that you don't. If you were to combine images before applying a light frame to each individual image, the light frames would no longer contain the correct values needed to offset the noise in your image.

Applying light frames after combining images could actually make things much worse. The same holds true for dark frames (if you didn't already have these subtracted during image capture) and for bias frames (if you need these because you are scaling your dark frames and applying them after image capture). With a one-shot color camera it is extremely important that you apply these calibration frames to the raw images prior to separating out the individual color components for image processing. If you were to separate the color components before applying calibration frames, noise gets interpolated out onto other pixels and, again, the calibration frames no longer contain the exact values needed to correct your images.

This point can't be stressed enough! Trying to apply calibration frames to individual color component images taken through a one-shot CCD camera the way most books tell you to do for color images taken through individual colored filters will most likely make a real mess of your final image.

The other reason to save each and every raw image versus just saving the combined image is that you will want to look through each individual image and weed out the bad ones. There are many reasons why an individual image can go sour and therefore be unusable in your final stack. Cosmic rays or a passing satellite or an airplane may have left a long visible streak on your image. Wind or a bump to your telescope or mount may have caused vibrations to show up in your images. Your autoguider (if you used one) could have had some trouble correcting for the movement of the telescope or the periodic error inherent in your worm gear. Any number of factors could rule out an individual image, but if you have taken multiple images and saved each one separately, then you can pull these faulty images out of the mix and still end up with enough good images leftover to combine them into a stunning final image.

Most image processing programs will allow you to open up all of your multiple images and work on them as a group. You can then easily scroll through each image individually and remove any images that are less than perfect when compared to the norm. After removing any bad images, you can then calibrate all of the images in the image group at the same time, applying light frames, dark frames, and bias frames, if needed, to each individual image with one command. After all of your acceptable images are calibrated, you will probably want to separate the individual color components. The imaging-control program that comes with one-shot color CCD cameras will have a simple command available to do just that.

The next step is to combine each set of color component images into a master one-color image. This process is fairly simple, but each image processing program has a little different routine, so you will need to read your instruction manual to understand the specific steps and options available in your particular program. The basic idea, however, is common across all imaging programs.

The first step in the combination process is to choose a point of reference that allows your image processing program to adjust each individual image so that the objects in the images will line up in the same pixels for each image in the stack. This is accomplished by choosing a couple of stars as reference points. The image processing program will calculate the centroids of each of these stars and then ensure that the centroids appear on the same pixels in the entire stack of images. These stars do not need to be the brightest stars in the image, but you do want to stay away from stars on the edges of the images so you can be reasonably sure that the same stars actually appear on each of the images. Choosing stars on opposite sides of the images will also allow for a more precise alignment rather than picking two stars that are right next to each other.

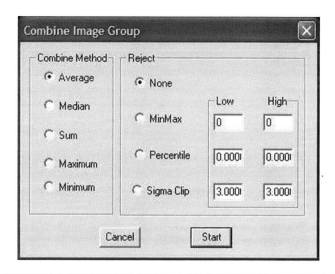

Fig. 9.8 There are several different mathematical calculations that can be used to combine multiple images

Choose stars that are well away from other nearby stars; if you are aligning images of a globular cluster you don't want the program to try to identify the exact star of a mass of tightly packed stellar objects. Also try to choose stars that are set in a dark sky area as opposed to stars that may be in an area with other objects such as nebulosity or the arms of a galaxy; the surrounding areas pixel values could throw off the calculations of the centroids. Once your images are properly aligned, then you can stack your multiple images into a combined image that will reduce noise and provide a better starting point for image processing and color combination than you would be able to achieve otherwise.

There are many different mathematical calculations that you can choose from in order to stack your multiple images. Most image processing programs will offer several of these combination methods along with options to modify the mathematical calculations in order to remove outlying data or noise from the calculations (Fig. 9.8).

In order to combine multiple images, the image processing program will take the value of each pixel location from each individual image, perform the chosen calculation on the data set of pixel values, and then assign the resulting value to that specific pixel location in the combined image. For instance, the image processing program will start by looking at the pixel values for each individual image in the pixel located at row 1; column 1. After determining the calculated value, the program will assign that value to the pixel located in row 1 and column 1 in the combined image. The program will then move onto the pixel values for each individual image in row 1; column 2, and calculate the value and assign this value to the pixel in row 1; column 2 in the combined image. The program will continue through each

Example of Average Combine Process	
Image	Pixel Value
1	1,534
2	1,222
3	12,314
4	11,970
5	14,913
6	18,444
7	11,580
8	12,430
9	13,434
Total Value	97,841
Avg. Value	10,871

Fig. 9.9 Example of pixel values from nine images used in an Average combine process

pixel location, determining the calculated values in each location and then assign the value to the pixels in the corresponding locations on the combined image...all in the blink of an eye.

The most commonly used mathematical calculations for the combination of astronomical images are an *Average combine* and a *Median combine.* An Average combine will take the values of all corresponding pixels, add them together, and then divide this total value by the number of images in the stack. However, this is not the most ideal way to combine images, as it allows noise and the values of other aberrations to be used in the calculation of the pixel value in the combined image.

For instance, let's say the values of the pixel located in row 5; column 25, show the following pixel values in a set of nine images (Fig. 9.9):

The pixel values in the first and second images show they are very dark compared to the other pixel values in this location. Perhaps these pixels fell in a shadow that didn't quite get cleaned up correctly when you applied your flat-field images. Also, the pixel value in the sixth image is very bright compared to the rest of the pixel values in that location. Perhaps this is the result of some random dark current that your dark frames didn't correct (it happens). Using an Average combine process, the total of these values are divided by the number of images in the stack (nine), and the program assigns a value of 10,871 to that pixel in the combined image. As you can see by looking at the values of the pixels on the list, most of the values fall in between 11,000 and 14,000, yet the final value that the Average combine process will assign to that pixel is outside of this range. Obviously the outlier values in the first two images and the sixth image have thrown the calculation of the "true" value off, and the results ended up at a less than optimum final value.

If a Median combine process is used instead of the Average combine process on the same set of pixel values, the result is quite different. In the Median combine

Example of Median Combine Process	
Image	Pixel Value
2	1,222
1	1,534
7	11,580
4	11,970
3	12,314
8	12,430
9	13,434
5	14,913
6	18,444
Total Value	97,841
Median Value	12,314

Fig. 9.10 Example of the same nine pixel values used in a Median combine process

process, the list of values is sorted from lowest to highest, and then the value in the exact middle of the list is used as the final value (Fig. 9.10). This process gives a final result of 12,314. This is a much better value than the Average combine process resulted in. It falls in between the approximate range of most of the values for this pixel. There is still some room for doubt, however, since there were two images that were noticeably darker at this pixel location and only one image that was noticeably brighter. This resulted in the middle image in the sorted list being lower than the "true" value of the pixel.

A better option for both of these combination methods is to use what is known as a "clip" to remove the outlier data points from each pixel's set of values (Fig. 9.8). You could use the clip in conjunction with the Average or a Median combination method, or any other method you may want to use to combine your multiple images.

There are several ways to identify these outliers, including assignment of a minimum and maximum value, assigning a percentage of the number of pixels, or by using a statistical reference point known as a Sigma Clip. By removing the outlier data points, in other words, minimizing the noise in the image data, the final result of each pixel's value becomes much closer to what the actual value should be.

In a min/max clip, you would assign the minimum value that the pixel must meet in order to be included in the data set for that pixel's calculation. Likewise, you would set the maximum value for which any pixel whose value exceeds the maximum allowed value also has its data point removed from the set of data used in pixel value calculation. Unfortunately, using this method assigns the same floor and ceiling to all of the pixels in the image, so it is not very useful in eliminating all but the very worst values of noisy data points.

Example of Combine Process with a 10% Min/Max Clip	
Image	Pixel Value
2	~~1,222~~
1	1,534
7	11,580
4	11,970
3	12,314
8	12,430
9	13,434
5	14,913
6	~~18,444~~
Total Value	78,175
Avg Value	11,168
Median Value	12,314

Fig. 9.11 Example of the same nine pixel values used with a Min/Max clip where some of the outlier data values are eliminated from the calculation

A better way to reduce noise in each pixel is by using a clip that eliminates a percentage of the minimum and maximum values for each pixel's data set. In this process, instead of assigning a set minimum and maximum value for all pixels in the image, you would assign the percentage of the data set that the minimum and maximum values for each individual pixel would remove.

Using the same example shown above, we can assign a 10% minimum and maximum pixel value for the given set of pixel data points. With nine images in the data set this clip then removes the lowest and the highest 10% of pixel values (one of each, 1,222 and 18,444) (Fig. 9.11). Having these outlier data points removed from the calculation provides an Average combination value of 11,168, but with the Median Combination value, because we removed the same number of outlier data points from the top and bottom, the value remains at 12,314.

An even better way to combine images is using a more complicated clip calculation known as a Sigma clip. This clip will calculate the standard deviation of each pixel as compared to the pixel values surrounding it. You would then assign a multiplication factor, such as three standard deviation values, that the image processing program will use to eliminate any values that fall outside of this range when compared to the values of the data set.

Using this process, the two darker pixel values and the value of the brighter pixel are completely removed from the calculation. Removing these outlier data points now results in a final average value of 12,774 and a final median value of 12,372, both of which are much closer to the "true" value of the data that should be represented at that pixel location (Fig. 9.12). The image processing program can perform this same

Example of Combine Process with a 3.0 Sigma Clip	
Image	Pixel Value
2	~~1,222~~
1	~~1,534~~
7	11,580
4	11,970
3	12,314
8	12,430
9	13,434
5	14,913
6	~~18,444~~
Total Value	76,641
Avg Value	12,774
Median Value	12,372

Fig. 9.12 The Sigma clip uses the statistical qualities of the data set to remove the outlier data values from the pixel value calculation

calculation for each individual pixel location. The Sigma clip offers the optimal solution because it uses the statistical qualities of the data set to automatically eliminate the outlying data points that are clearly incorrect. Eliminating this level of uncertainty results in a direct reduction of noise during the combination process, leaving your final image as good as it can be with the remainder of the data.

Other image combination methods are also available, such as a Sum function, which can be used when the data you are interested in has very low values when compared to the overall available dynamic range of the image. Under another combination method you can instruct the imaging control program to assign only the minimum or maximum values of each pixel's data set to the final image's pixel value. However, these methods can often include more noise than the other combination methods. Using a clip along with the minimum and maximum combination methods can help tame the noise factor if you need to use these combination methods to brighten or darken an image you are stacking.

Although taking longer exposures is almost always preferable to taking shorter ones, combining a series of shorter exposures can be almost as good. Even when using longer exposures, you will probably still want to take several images and combine them together. The reason for this is that combining images increases the signal-to-noise ratio for any set of images. When you combine multiple images, the signal value always increases faster than noise. Using the Sum combination method, the signal values increase along a linear function (the sum), while the noise only increases by the square root of the number of images.

For example, when combining four separate images all with an approximate signal value of 8,000 and a noise value of 200 (a signal-to-noise ratio of 40:1), the signal values are added together, resulting in a combined signal value of 32,000. The noise values, on the other hand, only increase by the square root of the number of images (the square root of $4 = 2$), so the new noise value ends up being 400 (200×2). This combination of images then results in a new signal-to-noise ratio of 32,000:400, or 80:1. By combining four images you have doubled the signal-to-noise ratio, resulting in a much clearer, sharper image with more contrast and a greater amount of detail present in the combined image.

The benefits of the increased signal-to-noise ratio are such that no matter how long of exposures you are taking you will almost always want to take several and stack them. Keep in mind, getting in the habit of taking multiple exposures of your targets also allows you to weed out any bad images that can't be salvaged, leaving you with at least some good images to work with.

Once you have your images stacked into one combined image, you can either process the image as is, or more likely you will combine the individual color component master images into a single full-color image for further image processing. Once the individual color component images are normalized and processed how you want them, you will recombine the individual red, green, and blue color components into an RGB image. You can also add what is known as a "luminance" image for an LRGB image, where the luminance image, an unfiltered full-spectrum image, is used to help bring out fine details in your recombined final image. Combining color images will be covered in much greater detail in the following chapter.

Chapter 10

Image Processing and Enhancement

Image processing is a very lengthy and complicated subject. An entire book could be written (and many have been) on the subject of image processing alone. Because of the depth of this subject, this book will only cover the basics of image processing, and it is suggested that you seek out as much additional information from other sources as you can. Image processing is as much an art as it is a science. Many of the techniques that can be used result in a subjective choice as to whether you like the results they have on your final image.

Certain image processing techniques tend to work better than others on images of different types of astronomical objects. For instance when imaging planets, a "sharpening" technique can bring out otherwise unseen details that are buried deep within the pixels of your image. When processing images of spiral galaxies, a "non-linear stretch" technique can help bring out details of both the cores of the galaxies as well as the galaxies' spiral arms.

The choice of which image processing techniques to use are yours and yours alone. There aren't any right answers or wrong answers when it comes to image processing; it all boils down to what kinds of details you want to show in your images and how you think the techniques have worked in order to bring out this detail. Be sure to save your original starting points and each of the successive variations of the image processing techniques so you can revert back to an earlier version if you don't like the way the last one turned out.

With most of the image processing techniques available, you can always try the process again using more or less of the values involved in the particular technique to see if that value works better for what you are trying to accomplish. Experiment, try different techniques, try different settings and values for these techniques;

L.A. Kennedy, *One-Shot Color Astronomical Imaging*, Patrick Moore's
Practical Astronomy Series, DOI 10.1007/978-1-4614-3247-0_10,
© Springer Science+Business Media New York 2012

eventually you will learn what works for you and what doesn't, which will save you time when processing future images.

Processing One-Shot Color Images

This section covers the most important step in imaging with a one-shot color CCD camera. If you learn nothing else from this book, you will want to learn the following fact about using a one-shot color CCD camera. It is absolutely critical, in order to achieve the optimal imaging quality from your color CCD camera, that you capture your images in the raw format as opposed to capturing only the stack of images or having the camera control software separate the images into its various red, blue, and green components prior to calibration. In order to get the best possible final results, you will want to begin your image processing steps from this starting point.

Virtually any book or article you may read about processing CCD images will state that you are to do most of your image processing on the red, blue, and green images separately. When taking images through separate colored filters this makes perfect sense, but based on the way that a color CCD camera transposes images into the colored components, you must perform several image processing steps *prior* to separating the individual color components for additional processing. Again, this is the most critical step involved in imaging with a one-shot color CCD camera and the most important piece of information you can learn from this book; perform the basic image processing techniques on your raw format images *before* you separate them into their individual color components.

Calibrating One-Shot Color Images

As was mentioned in the first chapter, color CCD cameras capture images using an array of pixels that alternate between red, blue, and green. In order to generate an entire image in one of the single color components, the data for each specific set of colored pixels are used and the data for that color is interpolated from the pixels that capture data in the other two colors. If you separate the individual color components and thereby perform this interpolation before removing any flaws or defects caused by your camera or telescope, these flaws will also get interpolated, making it impossible to properly remove them at this stage of the process.

The correct way to handle this issue is to calibrate your raw format images before you separate the individual color components and also before you perform any other image processing techniques on your images, including stacking multiple images. Starting with cleaned up, fully calibrated raw images is the best way to eliminate noise from the get-go, and doing so will pay huge dividends in the quality of your final images. This includes not only applying dark frames, which hopefully you already did at the time you were capturing your images, but also applying flat-field frames and bias frames if you are using those to scale up post-imaging dark frames. Following this procedure ensures that any flaws or defects are removed entirely and not interpolated unnecessarily into your color images.

Aligning and Stacking One-Shot Color Images

Once your calibration frames have been processed into your original raw images you can then begin stacking multiple images as outlined in the previous chapter. Again, you want to perform all calibration steps on the original raw images, prior to separating individual color components, so you are not interpolating flaws or defects in your images. If you separate your color components prior to calibrating your images, you will change the values associated with those flaws due to the separation process, thereby rendering your calibration images useless.

Additionally, the alignment process required as the first step in stacking multiple images can cause problems when trying to calibrate your images after separation. Even if you leave your images in the raw format without interpolating color data from the Bayer array, if you tried stacking your images prior to calibration, it is very likely that the flaws and defects inherent in your original images will line up on different pixels due to the alignment process, which is based on reference stars or other features of your targeted imaging subject. The alignment process will shift your images based on the pixel location of your reference stars. This will not only shift the stars as well as the imaging subject to the same pixels in all images, it will also shift any flaws and defects to different pixels on each frame. What you will end up with is a series of defects, such as hot or cold pixels showing up in a line in multiple pixels locations (Fig. 10.1). Doing so will once again render your calibration frames useless in performing the necessary cleanup that a high-quality image requires. Always calibrate raw images first and then align them before performing any other image processing or color separation procedures!

Once you have your raw images calibrated and aligned, you then need to decide if you want to convert your original raw images into their various color components for further processing. One scenario is to stack your raw, calibrated, aligned images into a combined image and then separate the individual color components from this master image. This can save you a good deal of time because you will only need to perform the stacking procedure once. This scenario can lead to decent final images, but it can also leave a bit of uncertainty or noise in the individual color components and also reduces the level of fine detail that you are able to bring out in your final image. The other scenario, which does involve quite a bit more time to do, is to separate the individual color components from each of your raw images first. You will then stack each set of color components into a combined master stack, which will then be used for combination into your final color image. In order to get the highest-quality final images, this scenario is highly recommended.

Separating the individual color components from your original raw images and then stacking each color component by itself will help reduce noise and uncertainty in each of the color channels, leaving a much smoother set of color components to

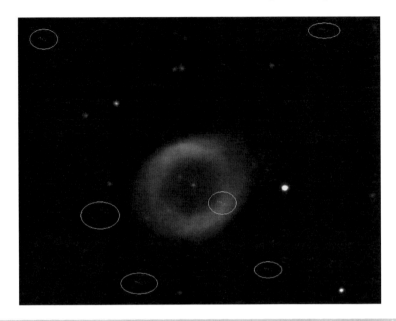

Fig. 10.1 Aligning raw images prior to calibration can leave a series of flaws that can no longer be cleaned up with your calibration frames

use when you combine them into your final color image. Because of the way the stacking process works, especially when performing the average or median combination with a clip to eliminate outlying data points, stacking each color component separately will not only do a better job of cleaning up the individual color channel but will also enhance the level of detail inherent in each color spectrum that can be drawn out of your images. The various colors can capture different specific pieces of data and detail from your targeted subject, and by performing this part of the image processing procedure on each color separately you can bring out the best parts of the various colored images before recombining them into your final image. Maximizing the individual color detail in each color channel before combining the color channels into a final color image will allow a more accurate representation of the targeted imaging subject and will also provide more details for you to work with when performing additional processing techniques on your final image.

Processing Individual Color Components

In order to convert your raw images into a final color image you will need to separate the color components out of the raw images, combine the individual color component stacks, perform some mild image processing, and then combine the color components into a full-color image. Although there are some image processing steps that you will want to make to the individual color components, such as stacking the individual raw color component images into a combined stack or normalizing the backgrounds of the combined color stacks, most of the image processing techniques that follow will need to be performed after combining the individual color components stacks into the complete color image. Many books and articles will talk about processing individual color components by themselves, but this relates to color component images taken through separate filters.

With color components derived from a one-shot color image it is highly recommended NOT to perform a lot of image processing on the individual colors. Doing so can severely throw the color balance out of whack, making it next to impossible to recombine the color components and have the true colors come out in the final image. When trying to do too much processing on individual color components, It is also possible to alter features of the image so they do not line up correctly on the same sets of pixels in the final image, again making it next to impossible to recombine the individual components correctly.

For instance if you perform some image processing techniques on the individual color components stacks, you may shrink the sizes of stars in one color more than in the other colors. When you then combine the individual color components into the final color image, you will notice that the fringes of stars show up in one color rather than having completely white stars that occur when all three color components are combined equally in each of the pixels that the stars cover. Colored fringes on stars are very difficult, although not impossible, to fix in your final image. In order to fix these, you will need to do some image processing work on each individual star; depending on the number of stars in your image, this can be an enormous undertaking. You are typically much better off combining the color components prior to performing a lot of image processing techniques.

On the other hand, there are certain types of image processing that you will want to perform on the individual color components prior to recombination. For instance, when processing images of planets, the need to sharpen these images in order to bring out the fine details that are hidden in the data is so great that you will want to sharpen each of the color components prior to combining the components into a final color image. Doing so can still throw the color balance way off and require a lot of time and effort to readjust the color components in order to bring the right color balance out in the combined color image, but the level of detail that will be brought out by performing this process is enough to make the extra time spent rebalancing the color well worth the effort. In virtually all other cases, you will want to combine the color components prior to doing any major image processing.

This is especially true of non-linear histogram stretching, which will be covered later in this chapter. Non-linear stretches are used to reallocate the desired data across the limited available range of the color spectrum that can be displayed in your images. This technique can work wonders on a combined color image, but if performed on one or more of the individual color components can throw the color balance so far out of order that it will be virtually impossible to combine the colors into an accurate and pleasing final image. Because of this fact it is highly recommended that you do not perform any non-linear stretching on the individual color components, but instead wait to perform these stretches on the final combined color image. Better yet, perform these stretches on a luminance image that you can combine with your combined red, green, and blue image in order to maintain the color balance between the individual color channels, yet still use the non-linear stretch to pull out and enhance the fine details in your final color image.

Combining Color Components

With your individual raw color-separated images aligned and stacked by color, you are now ready to combine these stacked images into your final color image. If you have chosen to perform any image processing on the individual color stacks, be aware that this can cause problems when combining the individual color components into the final color image. Otherwise, as long as you have aligned all of your raw images properly prior to separating the individual color components, you should have little trouble combining the individually stacked color components into a stunning combined final color image.

The human eye is an amazing piece of anatomy that utilizes different parts of the physical light spectrum to see the world around us in full and vibrant color. In order to accomplish this feat, the retina of the human eye contains different receptors that pick up colors in the red, blue, and green parts of the light spectrum. By combining these various colors in differing values, the eye can see any color in the rainbow. One-shot color CCD cameras are modeled after the human eye and therefore work in much the same way. These specialized CCD cameras also contain receptors, or sensors, that capture photons in the red, green, and blue portions of the light spectrum. The camera control software then combines these captured light spectrum values in their differing amounts to recreate all of the various colors that the human eye can perceive.

Just like you learned while doing your kindergarten art projects, combining two primary colors can give you a secondary color (red and blue makes purple, etc.) and combining two secondary colors together can give you a tertiary color. Just like in art class, varying the amount of the primary colors used determines how rich the secondary color turns out to be. Similarly, when the captured values for each of the color channel receptors are combined in a final color image, the image produced can recreate the full range of the color spectrum. Even better than in your kindergarten art class, when you use a CCD camera to combine equal values of all three colors, you end up with the pure white that is typical of how most of the stars in your images will be displayed. (In kindergarten, combining any amount of three colors usually just ended up being a mess.)

Most astronomical image processing programs will offer a processing routine that allows you to combine red, blue, and green images into a full-color image. This process may be a bit counterintuitive at first because you will notice that when you separate the individual color components from your raw images they don't come out actually looking red, green or blue, but instead, all three "colors" come out in grayscale. Don't let this fact fool you, however; what is being displayed in these separate grayscale "color" component images are the values recorded at each pixel location for that individual color. The image processing software will use these values during the color-image combination process to assign those of the given color component to that pixel location in the final combined image.

Before you begin the color combination process, you will want to perform an image processing step known as "normalizing" on your individual color component images. What normalizing does is to make sure that the background data (dark sky areas) have the same approximate value in all three color components. In theory the darkest areas of an image, where there is an absence of any light, should have a value of zero. In practice, however, due to the presence of light pollution as well as the noise inherent in the CCD camera itself, the darkest areas of your images will have a reading well above zero.

Unfortunately light pollution isn't spread out across the color spectrum evenly; for instance, sodium vapor lighting common in some street lights produces light pollution that falls in the green part of the spectrum. Mercury vapor lighting, also commonly used in street lamps, tends to produce light pollution that is heavily weighted in the orange part of the light spectrum (a light pollution suppression filter can block these sources of light-pollution). Because light pollution affects each color channel differently, the three individual color components will typically end up with varying levels of value for the background. Normalizing the three color component images will bring the background values to the same starting point, allowing the rest of the individual color data to be correctly spread across the dynamic range of the combined image. Failure to perform this step will make it extremely difficult to balance the colors in your final combined color image.

Most image processing programs have a feature available to open all three color component images and normalize them against each other. Although these programs will usually do an adequate job of normalizing the background data, you may find you have more precise control over this process by performing this function manually. In order to do this, you will need to open and measure the pixel values in the darkest areas of each individual color component image. After determining which color image has the lowest pixel values in the background data, you will then use a pixel math routine to subtract the difference in the background level values from the other two images. For example, if your red color component image has an average background value of 200 units and your green image has an average background value of 500 units, you will need to subtract 300 units of value from each pixel in the green image. Similarly if your blue color component image has an average background value of 300 units, you will then need to use the pixel math processing routine to subtract 100 units from each pixel value in the blue image in order to match the red and green background values.

Once you have your three individual color components normalized, you can then begin the combination process to merge the three individual color components into the final color image. Your image processing software will have a processing routine that will allow you to do this color combination. Although the names for this processing routine will vary between image processing software packages, such as "Join Colors" or "RGB Merge," they will all allow you the opportunity to select which image you want to use for each color component. Choose your stacked and normalized individual color component images for each appropriate channel – red image in the red channel, green image in the green channel, etc.

After you have selected the appropriate color image for each color channel, your image processing software will typically bring up a screen that shows the histogram of that individual color image, enabling you to set the black and white points. This provides the processing program the information on what the minimum and maximum values are for the range of data to be displayed in the final combined image for each color channel. More information on where to set these points will be covered later in this chapter.

It is important that you select the same approximate minimum and maximum values for all three color component images. This ensures that the data is spread out across the available dynamic range in the combined image in the correct way. If you were to have a larger range for one color component than the other two, this would tend to spread the displayed values for that color channel out more in the final image and will likely produce a final image with that color showing up more predominantly than the other two. Many programs will offer a feature that allows you to use the same min/max points that were used in the first image selected, or you can make note of what values the black and white points were set in the first selected color channel and then enter the data values for the other two color channels into the program manually when these images are selected.

Once the three color component images have been selected and the displayed value points are adjusted, your image processing software will typically show you a preview screen where you can make color balancing and other adjustments to the individual color channels prior to the color merge (Fig. 10.2). If for whatever reason

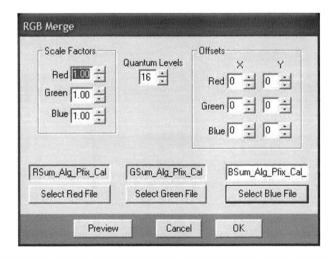

Fig. 10.2 The color combination program offers a number of adjustments that can be made to balance color and enhance the final color image

the color balance is off at this point, you can make adjustments to the color contribution for each individual color component. For instance, if sky pollution has left your image with a green tint even after normalizing the backgrounds of each color component, you can choose to reduce the color contribution of the green channel until the color balance is where you want it to be.

Another example of where this kind of adjustment might be needed is when you are imaging subjects near the horizon. Any time you are imaging targets below about 45° in the sky, a phenomenon known as atmospheric extinction will cause the data values in the blue spectrum to be reduced by anywhere from 10% to 40%. In this situation, you will need to increase the contribution value of the blue channel accordingly to compensate for the reduction in the values of the data captured in this area of the color spectrum.

Depending on your image processing program, you may also have the option of adjusting two other color values at this point – the color saturation and the color image gamma adjustments. Color saturation is a measure of the strength or intensity of the colors applied to the final color image. Increasing the saturation value will give you deeper and richer colors in your final image. The gamma adjustment determines how input values are correlated to the output values of a color's brightness. Think of it as a type of multiplier for brightness values; a setting of 1.0 will offer no change, but a setting of 2.0 will make the output values of the color data in the final image brighter. A gamma setting of 0.5 will, in turn, make the output values of the color data darker. Experimenting with the various settings of the color-merging routine will enable you to find the right balance of settings to make your final image look the way you want it. You can often find images of your targeted subjects available in books, magazines, or on the Internet, which will show you what your final images should generally look like. You can use these as a guide for your color combination to ensure the appropriate colors are represented correctly if color accuracy is the outcome you are shooting for.

Sometimes when you combine your color images, you will notice that all of the stars have some color fringes on one or more sides. If this happens, you made need to shift one or more of your color component images by a few pixels so that the images all line up correctly. Your image processing program should have a feature built into it that will allow you to perform this procedure. Other times, you may find some of the stars in your image are completely colored, as opposed to being white as they should be. In this case you will need a bit more drastic image processing. You may be able to readjust the color balance to minimize this color, thereby returning the stars to their natural white color. If that doesn't work there is a trick you can try, depending on what kind of image you are processing.

You should be able to select individual problem stars to fix, or if there are a lot of stars with problems you can use the Color Range selection tool in *Photoshop*™ to choose all the stars in your image. If your image has a lot of bright nebulosity, this may not work well as the Color Range tool will try to select the nebulosity with the same brightness values. In this case you may have no choice but to select and fix problem stars individually. If your imaging subject is confined to a small part of the screen, you can use the Polygonal Lasso selection tool in *Photoshop*™ to draw a "box" around it. Then use the "select inverse" option in the Select drop-down dialog box to exclude your image's subject and select the rest of the image instead. Once this is done, you can then use the Hue/Saturation tool to remove all of the color saturation from the problem stars. This method can also remove problem color splotches in the background as well.

Adding a Luminance Layer

In addition to a standard red, blue, and green (RGB) image, you can also create another type of color image known as an LRGB image. This type of color-combined image uses an additional component known as a luminance layer (the L in LRGB). A luminance image is a full-spectrum grayscale image that contains data from all of the wavelengths in the light spectrum. Adding a luminance layer to your RGB image allows you to make adjustments to the luminance layer, such as non-linear stretching, sharpening, and other image processing techniques that could mess up your color balancing if applied to an RGB color image. This allows you to get the best of both worlds – good color balance and enhanced fine details.

Most image processing software programs will create a luminance image from your raw images when you perform the color separation routine. If not, then it will probably allow you to create a luminance image from the individual red, green, and blue color component images. You will want to make sure to stack multiple luminance images just like you did with the individual color components in order to come up with a stacked and normalized "master" image to use in the LRGB combination.

Not all image processing software programs offer the ability to create an LRGB image, but these kinds of combined images can be easily created in *Photoshop*™. In order to create an LRGB image in *Photoshop*™ you will need to create a standard RGB image by opening the three individual color component images and applying them to individual channels in the channels dialog/RGB dialog box. You will then open the luminance image and convert it to "RGB Mode" using the Image/Mode dialog box and selecting RGB instead of Grayscale as the image mode. You will then copy the RGB color image that you created and paste it into the luminance image, which will then become a background layer.

Once the LRGB image has been combined, you will then need to adjust the "Opacity" percentage to allow both the luminance layer and the RGB channels to both be shown in the final image. A setting of 50% is usually a good starting point. Although adding a luminance layer can allow you to bring out fine details (adjusting it and then blending it with the RGB image to show both color and fine detail), there is a downside to using luminance layers. The downside is that the use of a luminance layer will tend to wash out the colors in the RGB image to some degree. This can be somewhat compensated for through the use of the opacity percentage. If you find that your colors appear too washed out with a 50% opacity setting, try raising the opacity setting in increments until the color comes back to an acceptable level.

With the LRGB combined image created, you will want to do the remaining image processing to the luminance image layer only and leave the RGB portion of the image untouched. This is especially true of non-linear histogram adjustments that would throw off the color balance of the RGB image. Any additional image processing can be performed on the luminance layer to extract as many of the fine details out of your image that is possible without fear of messing up the RGB color balancing. After additional processing of the luminance layer, you may need to make further adjustments of the opacity setting in order to allow the fine details that the luminance image holds to show up in your final image.

Processing Your Final Color Image

If you have created an LRGB image, then as stated above, you will want to do all additional image processing to the luminance image layer only. A luminance layer is completely optional, and as stated above, it can also have a downside by washing out the colors of the RGB image more than you are willing to accept. If you are not using a luminance layer then you will want to use any additional image processing techniques sparingly on your RGB image. Do your image processing in small increments and save your changes often so you can revert back to a previous version if the latest processing change has taken your image too far.

When saving your initial version of the combined color image, as well as subsequent changes to your color image, you will need to save them in a format that holds your color data. If you save them back to a FITS image, all color data is lost, and you end up once again with a grayscale image. For color images you will need to save them as TIFF files, JPEG files, Bitmaps, or some other format that allows the storage of color image data. Be very careful what kind of file format you save your color images into, as some image processing programs aren't designed to use certain file formats.

If you find that the desired image processing routines are taking your color balance too far out of whack to be fixable, there is another alternative. You can try stacking all of your raw images first without separating the individual color components and stacking them separately. You can then perform the desired image processing procedure to the raw stacked image, and when the image is processed just how you want it, you can then separate the individual color components from the raw stacked image. Using these individual color components, you can combine these red, green, and blue color components into an RGB image. Creating RGB color images in this fashion will contain a bit more color noise in the final image than you would have if you had stacked the individual color components separately, but this is fine as long as the additional fine details you want to draw out of your image are worth the trade-off. Keep in mind, there are no right or wrong ways to perform image processing; whatever method works for you to get your images looking how you want them is the right way to go.

Histogram Adjustment

In the previous chapter, it was mentioned that the histogram is a powerful tool used to define the data that is displayed on your computer screen during the image capture process. You will find that during image processing, the histogram again becomes one of the most powerful tools in your arsenal to make your images look the way you want them to. This applies not only to linear histogram adjustments, where you set the minimum and maximum data values, known as the black and white points, but also a technique known as a non-linear histogram stretch.

CCD cameras can capture a very wide range of different value levels. Most modern imaging cameras offer a range consisting of 65,000 different value levels. Unfortunately, the computer screens that the images are displayed upon typically only offer a range of 256 different value levels. If you were to allocate the 65,000 different value levels that the CCD camera captures across the 256 different value levels that are available on the computer screen, each level displayed on the screen would represent 253 different value levels of light with one shade of gray. Since many of the celestial targets that you will be imaging are captured with only several hundred or perhaps a few thousand of the 65,000 value levels, trying to display all 65,000 levels on the screen allocates the interesting detail of your targeted imaging subject to no more than a dozen shades of gray. The remaining 244 value levels will be allocated to different shades of dark-sky background and different shades of the brightness levels of individual stars – not the best use of 95% of the display values available on your computer screen.

In order to use the computer screen's display values to more effectively display your targeted imaging subject, you will use the minimum and maximum levels to define how much data to use in the allocation process. The histogram is the tool you will use to determine the right values in setting the minimum and maximum display values.

The histogram is a visual graph of the amount of data present in each pixel. Most histograms of astronomical images take the same basic shape, starting with a blank area on the far left of the histogram. This represents light pollution, or sky glow that causes the values of dark sky areas to read out at a value greater than zero. Next you will typically see a large spike on the left side of the histogram that represents the vast areas of dark-sky background, depending on your targeted subject (Fig. 10.3). This is followed by a sloping curve of data to the right of the spike. This is the area where the data for your imaging subject typically falls. Moving further to the right reveals a shelf of data that represent the brighter values of the stars in your image. This shelf will often run all the way to the right side of the histogram, depending on your image's subject.

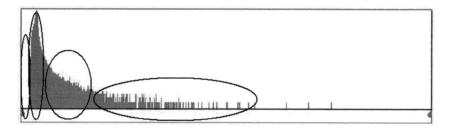

Fig. 10.3 The typical histogram for astronomical images has four major areas of interest, as shown here

As stated in the earlier chapter on histograms, the data for the dark-sky areas in your images should in theory read zero and appear all the way on the left side of the histogram. In practice, however, light pollution or sky glow will typically cause the values of dark-sky background areas to contain a value higher than zero. The more sky glow that is present, the higher these dark-sky background values will be. Since you will want to display the dark-sky background areas as black, or actually as a more eye-appealing dark grey, you will want to set the minimum value, or the black point, just to the left of the large spike on the histogram that displays the dark-sky background pixel values. Every pixel value below this minimum setting will now be displayed as black, hence the name black point. Most image processing programs will have a slider, or a small triangle that is used to set the minimum value on the histogram itself. Alternatively, you could type the minimum value directly into the appropriate box on the histogram control screen. Moving the black point slider to the right increases the minimum value point and darkens the image, especially the dark-sky background. Conversely, moving the black point slider to the left will lower the minimum value point and subsequently lightens the image.

Next you will want to set the maximum value level. Depending on where you set this level, every pixel with a value level higher than the maximum will simply be displayed in your image as a white pixel. Since these are usually the centers of stars anyway, it doesn't really matter that these pixel are displayed as white. To set the maximum level, or white point, you simply move the white point slider to the value level you want on the histogram, or, again, simply type the value directly into the appropriate value box on the histogram control screen. In similar fashion to the black point slider, moving the white point slider to the right increases the maximum value and darkens the image as a whole, while moving the white point slider to the left decreases the maximum value and ends up lightening the image as a whole. A comparison of a full scale image (Fig. 10.4) to the same image modified by using a linear histogram stretch (Fig. 10.5) shows the benefits of allocating more of the desired pixel values to the available display levels.

Setting the white point level is probably one of the most important steps in image processing. Since the values that remain between the black point setting and the white point setting are the values that will be allocated across the 256 available

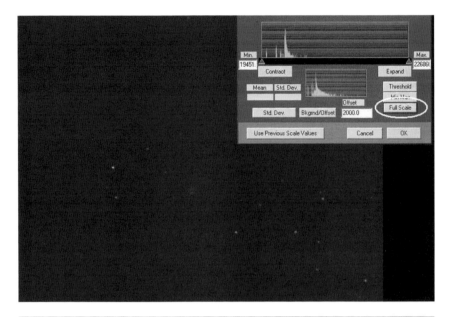

Fig. 10.4 A full scale image spreads all of the image's dynamic range across the limited display values on your computer screen and usually needs a histogram adjustment

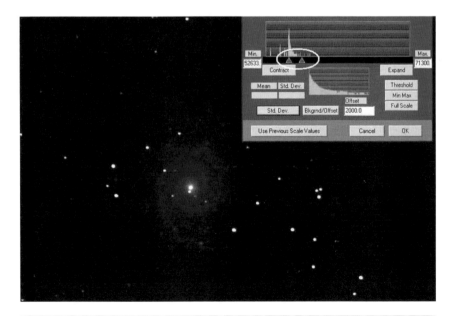

Fig. 10.5 A linear histogram stretch allocates the selected pixel data equally across the available display values on your computer screen

levels of your computer screen's display, where you set the white point is the biggest determinate of how many value levels will be allocated to displaying your imaging subject. When setting the white point slider, you will want to bring it as close to the sloping curve area of the histogram as you can without saturating or burning out any areas of your imaging subject. It is often a good idea to set this point slightly higher than where you might otherwise want it, to leave some room for additional image processing and possibly a non-linear histogram stretch.

Non-Linear Histogram Stretches

Many image processing programs will feature a third slider in the histogram tool known as a midpoint slider (Fig. 10.6). The use of this midpoint slider is the first blunt usage of a non-linear histogram stretch. Initially when you set the black point and white point sliders, the midpoint slider is spaced exactly halfway between these two points. This setting leaves all values between the black and white points equally allocated to the display values available on your computer screen. If you move the midpoint slider closer to the black point slider, the image processing program now allocates the smaller number of values between the midpoint slider and the black point slider to 50% of the available display values on the computer screen. This setting also allocates the larger number of values remaining between the midpoint slider and the white point slider to the other 50% of the display values available on the screen.

For example if there are 2,000 value levels between the black point and the white point, when equally allocated, one display level of gray on the computer screen would

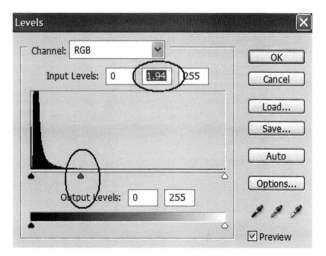

Fig. 10.6 When set off center, the midpoint slider allocates the pixel data from one side of the histogram across more of the available display values on your computer screen

represent about eight levels of pixel values in your image. If you were to move the midpoint slider to one-fourth of the distance between the black and white point sliders, the 500 values between the midpoint slider and the black point slider would be allocated to 50% of the 256 available display levels so that each computer display level for this set of pixel values would represent about four different levels of pixel values (500/128=3.9). The 1,500 pixel values between the midpoint slider and the white point slider would in turn be allocated to the 128 (50%) available display levels of the computer screen so that each display level would represent about 12 levels of pixel values (1,500/128=11.7). Instead of having the pixel levels allocated and displayed in a linear fashion (equal allocation), using the midpoint slider now allocates more pixel levels to one half of the available computer screen display levels and less pixel values to the other half – a non-linear stretch (unequal allocation).

Some image processing software will offer easy-to-use, preprogrammed versions of various non-linear stretches such as a dynamic non-linear stretch, which attempts to put the bulk of the actual image data into the available display levels based on the highest and lowest pixel values in the image, an exponential non-linear stretch which will tend to allocate more display levels to the darker pixel values and less display levels to the lighter pixel values which tends to darken your image, or a logarithmic non-linear stretch (Fig. 10.7), which will allocate more display levels to the lighter pixel values and less display levels to the darker pixel values, which in turn tends to lighten your overall image. You may also be able to adjust contrast and brightness levels in your images through the same control screen as in the examples below

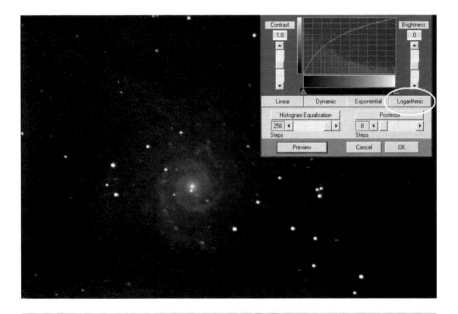

Fig. 10.7 A logarithmic non-linear histogram stretch will allocate more display levels to the brighter pixel data values

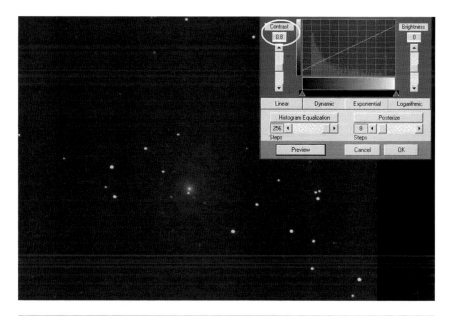

Fig. 10.8 Adjusting the contrast value of your image display can help highlight specific features of your targeted subject

(Figs. 10.8 and 10.9). Putting all of these tools into action together can provide a much better image than you started with (Fig. 10.10). Although these pre-programmed stretches can be adequate for your purposes and are definitely designed to be easy to use, you can get much more precise control of these and other functions through the use of the histogram routines and other processing tools built into *Photoshop*™.

In *Photoshop*™ you can use the Image/Adjustment/Curves dialog box to perform very precise non-linear histogram stretches on images that need to have their detail brought out further (Fig. 10.11). Using this tool you can add any number of anchor points along the histogram curve and adjust each section of the curve independently. This offers you much more precision in the implementation of your non-linear stretches and also allows you to avoid data clipping where the curves max out on the high or low end of the available display levels. Clipping artificially moves the extreme data values into the black or white point levels, causing them to be displayed as pure black or pure white. By assigning multiple points on the non-linear histogram stretch curve, you can control the ratio between the input value and the output value for each section of the curve. For the typical astronomical image, you will want to boost the output level for approximately the first one-third of the graph and then use additional points on the curve to keep the rest of the stretch from putting the slope of the curve all the way to the top. The last third of the curve should usually be a straight diagonal line (Fig. 10.12).

Photoshop™ also offers a wide range of other image processing tools, making this one of the most versatile and useful programs for astronomical image processing.

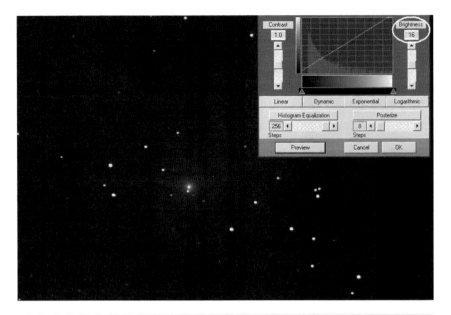

Fig. 10.9 Adjusting the brightness value of your images can make fine details in your images more or less visible

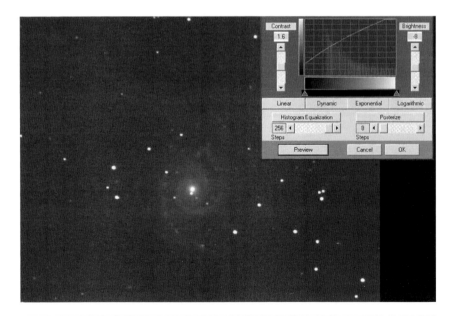

Fig. 10.10 The result of a linear and non-linear histogram stretch along with contrast and brightness adjustments (spiral galaxy M74)

Fig. 10.11 The Curves processing routine in *Photoshop*™ offers precise control over the shape and function of non-linear histogram stretches

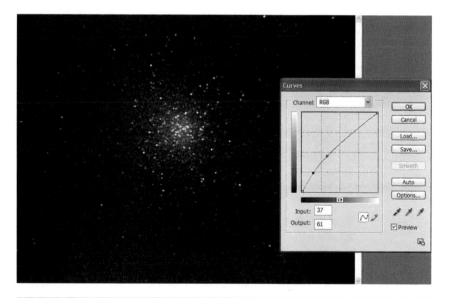

Fig. 10.12 Using the Curves tool, you can assign multiple points on the curve and adjust each separately to boost only those parts of the curve where your image needs enhancement

It is highly recommended that you spend a lot of time with *Photoshop*™ and discover all of the different options available for perfecting your images. In addition to the curves and levels tools, *Photoshop*™ also offers many features that other image processing programs don't, such as a Color Range selection tool, masking, layer building, pixel feathering, selection inversion, and many more that make this program indispensible for processing astronomical images. The time and effort you put into learning and using the many features available in *Photoshop*™ will pay dividends for your image processing skills and abilities.

Sharpening

Another useful and powerful tool that you can use to process astronomical images is known as sharpening. Just as the name implies, sharpening is used to make images that are slightly blurry or out of perfect focus more "sharp" or crisp. Sharpening can only do so much, and an image that is badly out of focus will probably just turn into a mess when sharpening is used. Noisy images are an even bigger problem for sharpening. The sharpening routine has no way to distinguish between noise and the details you would like to sharpen in your image. Instead, the sharpening routine will sharpen the noise right along with any details, generally leaving the image worse off than it was before you started. If you start with a good clean image with a low signal-to-noise ratio, or even just a fairly decent image, however, sharpening can make your images look better overall and can also pull out fine details that would otherwise not be visible in your final images (Fig. 10.13).

There are two basic kinds of sharpening routines. The first is brute sharpening, also known as high-pass filtering. This kind of sharpening works by increasing the brightness in contrast between various groups of pixels that are located adjacent to one another. You can set the size of the groups that the image processing routine will sharpen in your image, giving you some level of control over how much the routine sharpens as well as what gets sharpened. The smaller the pixel group size that you choose, the more detail in the image will be processed. A larger pixel group size will help keep noise from getting sharpened, too. Many image processing programs have a sharpening tool that includes a variable "degauss" filter that will also help eliminate noise sharpening.

You will need to experiment with the various settings that your image processing routine offers. You don't want to carry sharpening too far, or you can create false artifacts where there really aren't any. In addition, over-sharpening can tend to create "halos" or rings around the stars in your images (Fig. 10.14). Stars that are set against a light background, such as nebulosity or elliptical galaxies, can often end up with dark halos. Stars set against a dark background can also tend to end up with light halos when over-sharpening is used.

The other type of sharpening routine that is available with virtually all image processing programs is called "unsharp masking." Although it may sound a bit like an oxymoron, unsharp masking is an excellent routine to use to sharpen your images and bring out fine details in your targeted subjects. Basically, unsharp masking works by creating a "blurred" copy of the image and then subtracting the

Fig. 10.13 The sharpened image here shows much more detail than the original image in Fig. 10.11

Fig. 10.14 The effects of over-sharpening include artificial details from sharpened noise and halos around stars in the image

blurred, or unsharp, values from the original image, thereby leaving the remaining values as a sharper image than the original. The trick to unsharp masking is to use it sparingly – a little goes a long way.

Using an unsharp masking routine will also tend to sharpen noise along with actual image details. The image processing program has no way to tell whether differences between adjacent pixel values are legitimate, or simply noise from a non-legitimate source. For this reason you will ideally not want to use any kind of sharpening routine on noisy images. If you do need to use an unsharp masking routine on an image that is a bit noisy, you can experiment with the value setting of the routine to try and get as much detail as possible brought forth without enhancing the noise in the image. It may also be helpful to use a selection tool, such as the Color Range selection tool in *Photoshop*™ in order to select only the brightest parts of the image to be sharpened. Using this method you can exclude sharpening the background portions of the image, where noise can be most prevalent and sharpening can create false stars from noisy pixels.

Unsharp masking routines usually have settings for the amount of sharpening that will occur, the radius that will be sharpened (how many pixels away from the differing values are affected by the program's routine), and a threshold setting that determines how much of a difference in pixel values there must be for any sharpening to occur in that pixel location. If the sharpening routine ends up making the image too noisy, you can go back and try it again with a smaller radius setting until you find the one that works.

Just like with all image processing routines, you will want to experiment with the various setting to find one that works for that particular image. Remember, there is no right or wrong way to do image processing; the only thing that matters is getting the final image to look more like you want it to, no matter how you get there.

In most instances, when using a sharpening routine for processing your images, you will want to perform this step after you have combined your RGB or LRGB separate color components into the final combined image. However, because the use of sharpening is vitally important when processing lunar or planetary images, the sharpening routine should be performed on each of the stacked separate color component images before they are combined into the final color image. This still may tend to throw the color balance off, and it may take some extra effort on your part to rebalance the colors in the final image, but the added level of detail that you will be able to achieve by performing sharpening on the individual color stacks is well worth the added effort needed to combine the stacks after the fact. For virtually all other types of images, performing sharpening before color combination is a big no-no.

Smoothing or Blurring

After just discussing several ways to sharpen your images, the next image processing routine – "smoothing or blurring" – may seem like a step backwards. Why in the world would you want to make your images more blurry? Well, the answer is that you don't; you only want to make parts of your images more blurry – the noisy parts. In the section on sharpening, it was mentioned that the sharpening routine cannot distinguish between legitimate details of an image and unwanted noise that is also present. Sharpening the image, or using an unsharp mask, will tend to sharpen both legitimate details and the graininess of the noise in the image. Smoothing or blurring will help reduce the graininess that noise can present in an image, especially in the dark sky background areas.

Noise in an image, especially in the background areas, tends to manifest itself as random pixels that are lighter or darker than the surrounding pixels. If you get enough noise present, the sharpening routines can leave you with a speckled-looking background. This is why you don't want to sharpen noisy images; all of these noisy pixels tend to get sharpened, too, because there is a large difference in values between adjacent pixels. Even when the background of the image isn't sharpened, noise can still leave an unsightly appearance in your images, even more so if you have your black point set very high relative to the targeted subject's image details (Fig. 10.15). This is especially noticeable in color images, as the noise tends to show up in different colors. In order to help correct this phenomenon, you can blur the background and other noisy parts of the image so that all of the pixel values are more evenly matched in these areas.

Depending on the type of image you are processing, there are several ways to go about blurring the background. The simplest, when the subject is clearly different

Fig. 10.15 This image of the Cocoon Nebula (Caldwell 19) shows a lot of noise in the image, especially in the background

from the surrounding background portions, is to use the Select/Color Range dialog box in *Photoshop*™ to select just the background portion of the image for smoothing. Choose Quick Mask in the Selection Preview dropdown box so you can see what area of the image has been selected. Then use the eyedropper tool to point to a dark area of the background; this should select the entire background area and exclude areas that are brighter than this selection. Once the quick mask has been generated, you can then use one of the blur tools to process the area selected.

When the background noise has similar values to your targeted subject, you may need to use a more complex method (Fig. 10.16). In this image, the noise from the background blended right in to the edges of the nebula. Use of the Polygonal Lasso Tool in Photoshop allowed the nebula to be selected and then an Inverse Selection excluded the subject from the initial smoothing routine. Additionally, use of the Color Range Selection Tool allowed the stars and the other nebulosity in the rest of the image to be selected with a quick mask. By inverting the quick mask selection, all of the background portions were targeted for the smoothing process without affecting the subject or the stars in the image. By making another inverse selection and feathering the processing by 10–20 pixels, a different smoothing routine was used to take care of the core of the nebula.

The most common type of blur tool used for astronomical images is the Gaussian blur. Select the pixel radius for the smoothing to be spread out into, but don't go too high with this setting or you will create halos around stars as the edges get blurred out to the surrounding area of the background. In *Photoshop*™ you can use

Fig. 10.16 Using the Selection Tools to exclude the stars from the rest of the image, you can then use a smoothing routine to blur only the noisy portions of your images

the Preview option to see a real-time version of how the image will turn out when different values of the input variables are selected. After you complete the routine, if something doesn't look right, *Photoshop*™ also allows you to easily "step back" to a previous point in the process so you can try the routine again with different settings.

Once you have the background smoothed out, if you have also used a mask to sharpen the brighter details in your image, you can use the Layers feature in *Photoshop*™ to blend the two images. Simply open each image and use the Color Range selection tool to create a separate layer from each image. Create one layer from the smoothed background only and another layer from the sharpened parts of the other image. Paste both of these selections into separate layers and then use the Opacity setting, if needed, to allow both the sharpened part of the image and the smoothed part of the image to show through in the final combined image. Use the Layers/Group Layers tool to combine the layers together into the final image.

Digital Development

With the right images, the use of this next image processing routine, Digital Development, can be a real time saver. This is because the digital development process performs two functions at the same time. Digital development includes a sharpening routine coupled with a non-linear histogram stretch. This two-for-one approach is very powerful, but it can also be too powerful for some images. Color images don't fare too well when digitally developed, so it is best to use this processing routine on monochrome (black & white) images, or strictly on the luminance image if you will be combining an LRGB color image. If you are going to use the Digital Development routine to process your image, you will want to perform Digital Development before you do any other processing on your images.

Digital Development can work exceptionally well with images of galaxies and nebulae, enabling you to bring out the dim details inherent in the image as well as compressing the brightness range of the targeted subject so that the image is displayed better on your computer screen. Although this processing routine can be a real time saver for the right image, Digital Development can also be easily overdone. Because the use of sharpening is involved, the same problems that come from over-sharpening an image also come from the excess use of Digital Development. Noise in the image can be intensified, so noisy images are not very good candidates for the use of Digital Development. Similarly the overuse of Digital Development can leave halos around the stars in your image – dark halos around stars set against a bright background and bright halos around stars set against a dark background. Many times it is simply easier, and you can often gain more control over the process, if you just perform separate sharpening routines and non-linear histogram stretches on your images. But with the right image, one with a good signal-to-noise ratio, Digital Development can perform wonders in a very short amount of time (Fig. 10.17).

Fig. 10.17 Processing images with the Digital Development routine can save a lot of time and achieve excellent results with the right image. Compare this image of globular cluster M13 to the undeveloped version in Fig. 10.11

Digital Development works by running your image through a processing filter, also known as a "kernel." A kernel is best described as a type of mathematical matrix that tells the processing program exactly how to process the pixels surrounding an area of interest in your image. Most image processing programs offer several different types of filters that can be used for Digital Development. These are usually divided into low-pass and high-pass filters. Low-pass filters will tend to smooth images and block the processing of small differences in pixel values. In fact low-pass filtering can often lead to a reduction in the small levels of detail contained in an image. High-pass filters, on the other hand, tend to emphasize the small differences between adjacent pixels and ignore large-scale features. Determining which type of filter will work on a particular image is not always easy or intuitive, so it is probably best to try several of them and see which one works best through experimentation.

One of the more common types of filters used in Digital Development is known as a Fast Fourier Transform, or "FFT." This is usually a low-pass type of filter and will tend to give your image increased sharpening with less non-linear histogram stretching than other filter types. The FFT filter is often coupled with a setting that will allow you to choose the depth of the sharpening routine that is applied, sometimes referred to as the "hardness." You can often choose between mild sharpening, medium sharpening, or hard sharpening, or even create your own custom level if desired.

Another type of filter used in Digital Development is simply known as a Kernel filter. This type of filter is also typically available only as a low-pass filter. The Kernel filter provides a bit less sharpening and a bit more histogram stretch for your images than the FFT filter does. You will often find options with the Kernel filter to increase the level of sharpening that this tool will employ.

Again, the use of Digital Development can be problematic for noisy images and will probably turn those kinds of images into a real mess. If this is the case, then consider processing these images manually using a sharpening routine on the brighter portions of the image and a smoothing routine on the background portions of the image where noise is predominant. You can then perform the needed non-linear histogram stretch manually, giving you much more control over the final outcome than you will get by running the digital development routine. Digital Development also doesn't handle very bright images well, such as lunar or planetary images. Typically these types of bright subjects don't get much boost from the Digital Development routine and should be processed manually with sharpening and other processing routines in order to get the best results.

Deconvolution

The last type of image processing routine that this book will cover is known as "deconvolution." Deconvolution is yet another extremely powerful image processing tool that can be very easily overdone. Deconvolution can make a good image look great, or it can make an OK image look terrible! The deciding factor between these two outcomes is generally the signal-to-noise ratio involved. Images with a good signal-to-noise ratio, such as you get when taking long exposures, tend to be good candidates for the use of the deconvolution processing routine. Images with poor signal-to-noise ratios will not fare well with the use of this process.

Deconvolution is basically a method that can take the blurriness out of images. Blurring can occur for a number of reasons, whether it's from less than superb seeing conditions, defects or imperfections in your optical system, or perhaps from your telescope being ever so slightly out of the critical focus zone during your imaging session. Whatever the cause, deconvolution has the power to transform a slightly blurry image into a crisp, clear, sharp image in no time (Fig. 10.18). Deconvolution works through the use of a Point Spread Function, or "PSF." The Point Spread Function is a mathematical definition of how the light from a point source, such as a star, gets spread out across the pixels on your imaging chip.

The Point Spread Mathematical Function is basically a constant. In perfect seeing conditions, with perfect focus, through a perfect optical system, the light from a point source (star) would spread out across the pixels of an imaging chip in exactly the same way each and every time. Deconvolution looks at the point spread that has occurred in your image and works backwards from there to bring

Fig. 10.18 Deconvolution can turn a less than perfect image into something special

your image to the ideal point spread for all objects that your image contains. Just like other image processing routines, the deconvolution routine has no way to identify legitimate details from noise and will tend to create false features and artifacts in the backgrounds of noisy images. This is why starting with a good signal-to-noise ratio is so critical when using the deconvolution routine to process your images.

Deconvolution is another image processing routine that you must perform before attempting any other processing of your images. Deconvolution is often referred to as a "smart" sharpening routine because it doesn't just perform sharpening on whatever data the pixels happen to have in them, but it can actually correct data in pixels where there shouldn't be any, such as the excess parts of stars that show up as oblong in an image due to poor seeing, imperfect tracking, or for any other reason. The "smarts" inside the deconvolution processing routine can only do so much, however, and deconvolution, like sharpening and Digital Development, will tend to exaggerate any noise that is present in an image. For this reason, once again, noisy images are not good candidates for this processing routine. Here, too, it can't be stressed enough that taking the longest exposures you can is always best, since you will achieve a better signal-to-noise ratio in your images. Where long exposures aren't possible, and even when they are, capturing and stacking multiple exposures also helps improve signal-to-noise ratio and image quality. Using too much deconvolution can also create the dreaded false artifacts in noisy images and the light or dark halos around stars in images of good quality, so use deconvolution in moderation.

There are typically two main settings used in the deconvolution routine – the defined Point Spread Function (PSF) and the number of iterations that the process will run through. Unlike other image processing routines, deconvolution needs to work on an image several times in a row before you will see improvement in the image clarity. Many times, in fact, you will notice that the first few iterations of the program, the number of times it runs, will seem like a step backwards. As the image is deconvolved more and more with subsequent iterations, however, you will start to see the sharpness come out in the image. You will probably need to experiment once again with these settings to find the right number of iterations for the image being processed. Too few iterations and the image doesn't get processed as well as it could be; too many and you will start to see halos, artifacts, and other problems caused by too much processing. The correct number of iterations can be as few as 7 or 8 and as many as 20, 30, or even more. You will have to experiment with the iteration setting to determine how many iterations work on that particular image.

The other setting, determining the Point Spread Function, is a matter of using a Gaussian PSF, choosing a star for the deconvolution to calculate the image's inherent PSF, or indicating manually what PSF the program should use. This manual assignment is known as the "sigma" pixel value and indicates how many pixels the ideal point would be spread across for the given image. You should assign a small sigma – less than a value of 1.0 – if the image was captured with superb seeing conditions. If the seeing conditions were pretty good, but not superb, then you will need to assign a sigma value somewhere between 1.0 and 1.2. If the seeing conditions were merely average during image capture, then you may need to assign a sigma value all the way up to about 1.5; with images captured under poor seeing conditions, deconvolution may not be able to fix the image at all. If you assign too large of a sigma value for a given image you will start to see artifacts and halos showing up. If this happens, assign a smaller sigma value and try running the deconvolution routine again.

There are many different types of deconvolution routines that are frequently used for processing astronomical images. The most common ones are known as Lucy-Richardson (Fig. 10.19), Van Cittert, and Maximum Entropy. Each version of the deconvolution routine offers slightly different types of processing. Determining which one to use on a given image is largely a matter of trial and error. Be sure to save your original image and run different versions and different numbers of iterations off of the original image and then compare them to see which combination works best for the particular image that you are processing. When you have taken the deconvolution process as far as it can go, you can then try other processing routines to further enhance your image. Many CCD images can benefit from using digital development after the deconvolution routine has been run.

Fig. 10.19 Forty-eight iterations using a Lucy-Richardson deconvolution routine has sharpened the stars in this image and brought out more of the faint details within the nebulosity (Iris Nebula, Caldwell 4)

Summary

There are many other types of image processing routines, but for the most part they will simply be different variations of the various routines laid out in this chapter. Use whatever versions work best for your particular image. Remember, there aren't any right ways or wrong ways to process images. The only thing that matters is that the image turns out the way you think is best. Experiment with different settings and different processing routines to find the ones that work best for you. Don't forget to save your processed images often so you can revert back to a previous version if the latest setting or routine you used has taken your image into an area where you didn't want it to go. Most importantly, keep in mind that image processing cannot necessarily make a bad image good, but it *can* make a good image great, so always work to improve your imaging skills!

Chapter 11

Displaying Your Images and Other Possibilities

Displaying Your Images

Once you have your images processed into their final version, if you're like most people, the next thing you will probably want to do is to figure out a way to show them off. Of course you can always pull them up on your computer and show them to people, but this can become a bit cumbersome and you would have to have your computer with you whenever you wanted to show them off. There are several other ways you can display your images, enabling you to share your passion with friends, family, and anyone else who may be interested.

If you want to show the images on your computer, one of the best ways is to save the images into a slideshow. This way you can boot up the presentation and run through all of the images for your captive audience. Another method that works well is to convert your images into a screen saver program or even use them as your background on your computer screen or cell phone. The smaller the screen, however, the smaller your images will be displayed, and oftentimes your images will not show the level of detail that you would like them to, which is generally what makes them so impressive to look at.

Another cool way to display your images is in a digital photo frame. These digital frames are available in various sizes, and with a large enough picture frame you can display an excellent level of detail with this medium. These digital frames will automatically scroll through a series of images so you can download as many images as your digital frame will hold and then set it to scroll through them at various time intervals. These digital frames can then be displayed anywhere around the house, your office, or any other location where you want to display your images.

L.A. Kennedy, *One-Shot Color Astronomical Imaging*, Patrick Moore's
Practical Astronomy Series, DOI 10.1007/978-1-4614-3247-0_11,
© Springer Science+Business Media New York 2012

Perhaps one of the best ways to show off your images is to upload them onto your own personal website. You can have your own personal website fairly inexpensively nowadays, and many times you might even have some free website space available through your e-mail provider. Once you have a website available you can upload as many images as you have space for and arrange them in any way you want to have them. With a little website development and maintenance you can display your images in an unlimited number of interesting ways. You can even add descriptions or captions under your images so people have a better idea of what they are looking at. Most people don't know that much about galaxies, nebulae, or other celestial objects, so a short description coupled with some interesting facts can be very entertaining for most people. On your website you can also explain the process of taking these images and let people know just how much effort went into capturing and processing them. Most people find this kind of information fascinating, and they will be extremely impressed with your hobby as well as your imaging abilities and technical know-how.

In order to get the word out, you can send out a link to your website and let people get there with a simple click of their mouse. Another great way to spread the word is to have some business cards made up or to make your own cards out of some thick paper stock that has the URL address to your website printed on them. You can then hand these out to anyone you would like to have access to your site and to see your handiwork.

Another awesome way to show off your images is to have prints made up. If you are going down this route, there are a few things you will want to be aware of. Most times the folks who process standard photos for a living don't really have any idea what they are looking at when it comes to astronomical images. You may want to provide a visual example, or at the very least, provide specific directions on colors and other image information, so they don't try to "fix" the images they are developing. Prints can be ordered in any size and can even be framed for display on your wall, desk, or in other areas where you want to show off your best and most interesting images.

One thing to keep in mind when deciding how to display your images is that bigger is not always better. If your images are "blown up" to be much larger relative to the original size, you can actually start to see noise and graininess that you wouldn't ordinarily see in the original size display. Some noisier images can even benefit from being re-sized into a smaller version that helps hide some of the noise and graininess that would otherwise be present in a full-sized image. When blowing images up beyond their originally intended size you will want to make sure you are choosing the highest quality images you have captured and that you are also using all of the available image processing techniques at your disposal in order to produce the best possible final images that you can.

You may also want to consider keeping a digital scrapbook in sequential order based on when you captured the specific images. Over time as you master your hobby you will be amazed at the increases in the quality level of the images you will learn to capture, as well as your increased ability to process these improved images into their final version. Practice makes perfect, and CCD imaging and image processing is no different in this respect. The more time you spend capturing images, the higher quality images you will be able to capture, and the more images you process, the better you will become at image processing. What at first seemed like a laborious routine, such as getting good polar alignment, or stacking multiple images in each separate color, will soon become second nature. You will eventually learn to fine tune all aspects of your CCD imaging routines along with becoming more familiar with your existing equipment and its capabilities. In time you will probably want to obtain even better equipment, which can also help improve your imaging. All of these factors will help push you along the learning curve faster and faster, and keeping track of your progress through a digital scrapbook will chronicle all of the steps you have taken along the way.

Eventually, you may be able to hone your craft to such a degree that you can submit your images to the ultimate display vehicle – inside the pages of astronomy magazines! There are also websites for these magazines where high-quality amateur astronomers' images are displayed. Equipment manufacturers are also always looking for high-quality images as examples of what their gear can do. Who knows, maybe some of your images will end up in the next ad for the latest and greatest CCD equipment or astronomical imaging software.

Whatever your goals are and wherever your images are eventually displayed, the biggest thrill will ultimately come from your own sense of pride and satisfaction at having mastered a difficult hobby – and that's all it will take to keep you coming back for more!

Narrowband Filter Imaging

CCD imaging cameras are very versatile. In addition to taking individual "true color" images of celestial objects in the night sky, there are many other ways that you can use your CCD camera. One option that has become very popular recently is to take "false color" images through specialized narrowband filters that "see" certain types of emissions from celestial objects, particularly in nebulae.

Narrowband filters are named for the types of emissions that they will image by allowing only the wavelengths of light for certain emissions to come through the filter. Hydrogen-alpha (Hα) is one interesting emission that occurs when a hydrogen atom has a jump in its energy level from absorbing photons from nearby stars. Triply ionized oxygen (OIII), sulfur (SII), and hydrogen-beta (Hβ) are other types of emissions that have their own filters specifically designed to pass only these wavelengths of light. By using these narrowband filters when imaging you can capture and display a completely different and unique perspective of well-known objects.

In order to do this kind of imaging, you will need to obtain one or more of these narrowband filters and drop them into your imaging train before and during imaging. Because you are trying to capture emission data, as opposed to color data, you will need to set your one-shot color camera so it will capture images in monochrome (black and white). This is accomplished with a setting in your imaging-control program. Imaging in black and white will allow your CCD camera to use its full dynamic range for the specific wavelengths of light you are trying to capture. This will also avoid the need for the imaging-control program to have to interpolate data from the Bayer array, thereby providing more accurate data and higher resolution for your narrowband images.

Another interesting idea, if you have a complete set of these narrowband filters, is to take separate images through three different kinds of narrowband filters, and then assign each type of image into a separate color channel to create a combined image of various emissions. For instance you can assign an Hα image to the red channel, an OIII image to the green channel, and an SII image to the blue channel. Process these images in the same way you would for a standard RGB combination, and you will have a very interesting perspective of your imaging subject, with the red, green, and blue colors showing the various emissions coming from your subject.

You take these images the same way separate colored filter images are taken. Ideally, you will want to use a "filter wheel" for your imaging train. A filter wheel is a piece of imaging equipment that allows you to load the various filters in a circular or rectangular holder so that you can easily switch filters between images without having to tear down your equipment. If you get a set of filters that are parfocal, or have the same focusing characteristics, you will not even have to adjust focusing between shots. If your filters are not parfocal, then not only will you need to re-focus your telescope for each filter, you will also need to capture light frames for each filter before you adjust focusing so you can calibrate each set of images correctly.

Solar Imaging

Instead of just imaging night-sky objects, your CCD camera can also be used to image the Sun during the day. In order to accomplish this you will need a specialized type of filter known as a solar filter, which will drastically cut down the amount of light coming through your telescope so the Sun can be safely imaged without destroying your CCD camera. You will almost always want to install your solar filter on the end of your telescope, before you begin installing your CCD imager or even pointing your telescope towards the Sun. Because of the Sun's proximity and the corresponding brilliance of its light, the light can burn out your CCD imaging chip in short order. The heat from the Sun can also cause damage to your CCD camera or your telescope if the solar filter is not installed before you point your equipment anywhere near the Sun.

***** **Warning** *****

NEVER, EVER look at the Sun through your telescope's eyepiece. Everyone knows that staring at the Sun can cause blindness, but due to the magnification power of a telescope and the manner in which it focuses the light into the eyepiece, looking at the Sun through a telescope, even for a fraction of a second, can cause *instantaneous* blindness!

****

Once your solar filter is in place, you would image the Sun and process the images in the same way you would image and process any other target. With a little luck, depending on the current solar activity taking place at the time, you can capture solar flares, sunspots, or maybe even a passing planet (Mercury or Venus) if they happen to be in the right position at the time you are imaging. You can get data for these planetary transits off the Internet, which will help you to plan your imaging sessions accordingly.

Imaging Manmade Objects

Most CCD imagers today are powerful enough that you can not only image bright celestial objects, you can also image manmade objects such as the numerous satellites that orbit Earth every day. Coordinates for most satellites, with the exception of the military's classified versions, are readily available on the Internet. You will need to load these coordinates into your telescope or imaging-control program, so your telescope can be pointed in the right direction when these objects are overhead.

Set your auto-guiding equipment to use these objects for tracking purposes so that, instead of centering on a relatively slow-moving guide star, the auto-guider will move the telescope very quickly to follow the movement of the satellite or other manmade object. Your timing will need to be just right in order to do this kind of imaging, as your imager will not be able to pick up the manmade lighting on these objects in space, but will instead need to pick up their reflections of sunlight as they pass by at dusk in the evening sky. With a little planning you can even get an image of the International Space Station as it orbits overhead!

Building Mosaics

Another interesting thing you can do with your CCD camera is to take multiple overlapping images across different areas of the sky (Figs. 11.1 and 11.2) and then combine them in an image processing program such as *Photoshop™* in order to create a very large combined image called a "mosaic." Some camera control programs offer functionality to capture mosaic images easily by defining the coordinates you need to take the images at and even positioning your telescope and imager to the predetermined coordinates after you take sets of images in each location.

You will need to adjust the histogram for each set of images to make sure that the backgrounds match each other in brightness and contrast. You may have to make some additional adjustments to the brightness and color of the border areas where the images overlap. You will want to make sure there is a smooth, unnoticeable seam between the images. The remainder of the image processing should be accomplished after the images have been combined into the mosaic.

Fig. 11.1 An image of the lower left side of the Eastern Veil Nebula

Fig. 11.2 A slightly overlapping image of the upper right portion of the Eastern Veil Nebula

Fig. 11.3 The two overlapping images of the Eastern Veil (C33) are combined into the start of a complete mosaic image

Leave some room on your white point setting of each image in order to do the final histogram adjustment to the mosaic along with whatever other image processing techniques you wish to employ. Combining the separate images into the mosaic will involve the use of the Layers function in *Photoshop*™ or a similar function in other image processing software. You will need to manually align the images on top of each other so that key features, typically a pair of stars, line up on top of each other when the images are positioned. In addition to vertical and horizontal alignment, you may also need to rotate the images to align them if field rotation has occurred during image capture or if you have taken sets of images on different nights. *Photoshop*™ offers tools to perform these functions very accurately.

Once the mosaic is created, you will save them all into one combined image (Fig. 11.3) and then perform whatever final image processing steps you want to in order to bring out the best details of the mosaic. Using this technique, you can image larger portions of the sky or image subjects that are much bigger than your image scale would otherwise allow.

Time Lapse Imaging

Many imaging-control programs offer ways to take time lapse images so that you can actually create a "movie" of interesting celestial phenomenon, such as the relatively fast rotation of the planet Jupiter. Using the program you will assign a specific time interval in which the control program will instruct the CCD imager to take a set of multiple images. After processing these images in the normal manner, you can use the imaging-control program to load these images into a series of frames, just like the frames of a motion picture film. When played back, the control program will set the time lapse images into motion, providing you with a short film of your targeted imaging subject. Follow the instructions that came with your imaging-control program to perform this image capture and time lapse creation.

Astrometry and Photometry

Besides just using your CCD camera to take color images of various night-sky objects, there are also a number of ways you can use your imager to do scientific "research." Due to the proliferation of excellent amateur researchers and the accuracy of measurements being taken nowadays, many scientific institutions will now accept amateur astronomers' measurements and imaging results in order to maximize their research efforts.

One area of research you can look at doing with your CCD imager is known as "astrometry." Astrometry is defined as the determination of the coordinates of celestial objects. Another area of scientific research that you can look at performing with your CCD camera is to take photometry measurements. This is a measurement of how much light is coming from a specific star or other object, also known as the "magnitude" of the object. By taking careful measurements from your imager you can create a graph of an object's brightness, known as a light curve, over the course of several hours, nights, weeks, or even longer if needed. For objects such as variable stars and eclipsing stars (where two binary stars pass in front of each other from our point of view), this light curve can provide very useful and interesting data.

Searching for Novae and Supernovae

Astrometric and photometric measurements can also be used in combination to image a number of interesting objects in the night sky. By taking images of the same area of the sky over the course of several nights, weeks, or months, you can measure the light output coming from variable stars and eclipsing stars; with a little luck you may even be able to find a nova or even a brand new supernova. These images, taken over the course of a long period of time, can be compared to each other, looking for obvious changes in the magnitudes of individual stars.

Most imaging-control programs offer features that you can use to compare images, as well as tools to perform the brightness measurements in order to identify the magnitude changes of variable stars, eclipsing stars, or exploding stars (nova and supernovae). The measurement tool will need to be assigned reference stars, whose magnitude does not change, so the magnitude shifts can be accurately measured. This eliminates any differences in brightness that can occur from variables such as poor seeing, light pollution effects, and specific image resolution. The magnitude data can then be loaded into a graphing tool, which will produce a graph of your subject's magnitude shifts, known as a light curve. Properly measured light curves can be uploaded to astronomical research institutions such as AAVSO, the American Association of Variable Star Observers.

Hunting for Asteroids, Comets, and Dwarf Planets

Similarly, your imager can be used for seeking out known or even previously undiscovered asteroids, meteors, comets, or dwarf planets. Many amateur astronomers dream about making a discovery of this sort, and you can easily perform searches of the night sky using your telescope and CCD camera.

In order to obtain images for these kinds of searches, you will need to take at least three images at different time intervals. How long these intervals should be depends on how fast the object you are imaging is moving. Most dwarf planets, for instance, have an apparent movement of between approximately 0.25 and 1 arcseconds per minute. Depending on the image scale that your CCD equipment is set up for, you will need to take images at least 10–45 min apart in order to have the object move across a number of different pixels on your imaging chip.

You take at least three images of the object in order to verify that the object is moving in a straight line. Two images can define the points of a straight line, and the third image can be used to verify that the object is in fact following the path of this straight line. This three-position measurement can help eliminate false sources of object data such as those generated by hot pixels or by other noise on your imaging chip.

Once you have these images captured, you will need to align these images based on some reference stars so that all of the "non-moving" objects will be in the same position. With the images aligned, you can then load these images into a "blinking" tool, which you will use to quickly flash these images on your computer screen back and forth, one after the other, in order to visually compare them. You will be looking for objects that appear in different locations on the images, and as stated they will need to be moving in a straight line across the images in order for them to be considered a moving astronomical object.

After you have identified a moving target, you will use the astrometry tools in your imaging-control program or your image processing program to calculate the exact right ascension (RA) and declination (Dec) measurements for your subject. To do this, you will identify for the measurement tool the names or the star catalog numbers for a couple of the known stars that also appear in your target's image. You will also need to indicate the image scale that you were using when you captured your images. The tool will then use the known coordinates of the reference stars and the measured pixel distance of the target from these stars, multiplied by the arc-seconds per pixel that your image scale is calculated at, to determine the coordinates of your targeted subject.

You will also need to calculate the apparent speed of motion for this object based on the time between the images you have captured, the pixel distance that the object moved, and the image scale at which your images were captured. Once you have this information, you can submit the discovery to an amateur astronomical association, such as the Minor Planet Center. Using the supplied data of the three coordinates and the relative speed, the trajectory can be calculated so the object can be found on subsequent evenings to confirm the discovery. The amateur astronomy association will also compare this information to the coordinates and trajectories of previously identified objects to see if this is a new discovery or not. If it is, in fact, a new discovery and you are the first one to enter a submission, then you may even get to name the object, depending on what the rules are for the organization responsible for the naming conventions of celestial objects, the International Astronomical Union (IAU).

Additional Resources and Star Catalogs

Resources

There is a wealth of information on astronomy, CCD imaging, and image processing available. These subjects are so in depth that there is always more to learn. Couple this with the fact that new discoveries are constantly being made in the field of astronomy, new equipment and hardware is being introduced, and new techniques for improving images through the use of creative image processing are being developed, and you will realize that this hobby has a learning curve that is always growing. Make it a point to seek out new resources and sources of information to help you improve your understanding and capabilities and you will always find new and interesting targets to image along with new ways to improve the imaging skills you need to capture, process, and display these targets.

Listed below are some of the resources you might use to advance your skills. New information is also being put on the Internet each and every day, so seek it out, save the best sites you find to your favorites folder, and check back often to see what new information is available for you to improve your skills.

L.A. Kennedy, *One-Shot Color Astronomical Imaging*, Patrick Moore's
Practical Astronomy Series, DOI 10.1007/978-1-4614-3247-0,
© Springer Science+Business Media New York 2012

Magazines

Astronomy/Kalmbach Publishing Co. (www.astronomy.com)
Sky & Telescope/New Track Media Co. (www.skyandtelescope.com)

Both of these magazines are excellent resources for the latest information on astronomy, telescopes, and related equipment, as well as CCD imaging and image processing techniques.

Books

The New CCD Astronomy by Ron Wodaski, New Astronomy Press, 2002, Duvall, WI

This book is an excellent resource for technical explanations of image processing. The book typically comes with a free image processing program for Windows-based computers.

AIP 4 Win by Richard Berry & James Burnell, Willman-Bell, Inc., 2005, Richmond, VA

This is probably one of the most extensive books on CCD imaging available on the market today. A must have for CCD imagers at all skill levels!

Sir Patrick Moore Practical Astronomy Series by various authors, Springer Publishing

This world-renowned series of books is the most extensive collection of works on all aspects of astronomy, CCD imaging, and image processing. Each and every book in this series will teach the reader something new regarding these subjects.

Websites

Digital Space Images/www.digitalspaceimages.com

This is the author's website, which contains astronomical images, CCD imaging equipment, and other helpful information regarding CCD imaging:

Meade Instruments/www.meade.com
Orion/www.telescope.com/orion
Celestron Telescopes/www.celestron.com

Websites for the manufacturers of high-quality telescopes, CCD imagers, and other equipment for astronomy and astrophotography:

Clear Dark Sky/http://cleardarksky.com
Starizona/www.starizona.com
Starry Wonders Astrophotography/www.starrywonders.com
Observational Astronomy/www.stargazing.net/david/misc/contents.html

Waid Observatory/www.waid-observatory.com
International Dark Sky Association/www.darksky.org

Websites with useful information of all kinds regarding CCD imaging and / or image processing.

American Association of Variable Star Observers/www.aavso.org
Minor Planet Center/www.minorplanetcenter.net/iau/mpc.html
International Astronomical Union/http://iau.org

Websites for the submission of measurements made by amateur astronomers:

Astronomy Picture of the Day/http://apod.nasa.gov/astropix.html
Advanced Observing Program – Kitt Peak Observatory/www.noao.edu/outreach/aop/

Websites with images from space and land-based astronomical observatories.

Jim Burnell's Homepage/www.jburnell.com
Robert Gendler Website/www.robgendlerastropics.com
Cosmotography by R Jay GaBany/www.cosmotography.com
Russell Croman Astrophotography/www.rc-astro.com

Websites from world-class CCD astrophotographers.

The Messier list

Messier number	NGC number	Object name	Type of object	Dimensions	Constellation	Apparent magnitude	Distance (K LY)
M1	1952	Crab nebula	Supernova remnant	420" × 290"	Taurus	9	6.3
M2	7089		Globular cluster	16'	Aquarius	7.5	36
M3	5272		Globular cluster r	18'	Canes Venatici	7	31
M4	6121		Globular cluster	36'	Scorpius	7.5	7
M5	5904		Globular cluster	23'	Serpens	7	23
M6	6405	Butterfly cluster	Open cluster	25'	Scorpius	4.5	2
M7	6475	Ptolemy cluster	Open cluster	80'	Scorpius	3.5	1
M8	6523	Lagoon nebula	Cluster with nebula	90' × 40'	Sagittarius	6	6.5
M9	6333		Globular cluster	12'	Ophiuchus	9	26
M10	6254		Globular cluster	20'	Ophiuchus	7.5	13
M11	6705	Wild Duck cluster	Open cluster	14'	Scutum	7	6
M12	6218		Globular cluster	16'	Ophiuchus	8	18
M13	6205		Globular cluster	20'	Hercules	5.8	22
M14	6402		Globular cluster	11'	Ophiuchus	9.5	27
M15	7078		Globular cluster	18'	Pegasus	7.5	33
M16	6611	Eagle nebula	Cluster with H II region	7'	Serpens	6.5	7
M17	6618	Omega, Swan, Horseshoe or Lobster nebula	Cluster with H II region	11'	Sagittarius	6	5
M18	6613		Open cluster	9'	Sagittarius	8	6
M19	6273		Globular cluster	17'	Ophiuchus	8.5	27
M20	6514	Trifid nebula	Cluster with H II Region	28'	Sagittarius	6.3	5.2
M21	6531		Open cluster	13'	Sagittarius	7	3
M22	6656	Sagittarius cluster	Globular cluster	32'	Sagittarius	5.1	10
M23	6494		Open cluster	27'	Sagittarius	6	4.5

	NGC	Common name	Type	Size	Constellation	Magnitude	Distance
M24	4715	Sagittarius Star Cloud	Milky Way Star Cloud	90'	Sagittarius	4.6	2
M25	4725		Open cluster	32'	Sagittarius	4.9	5
M26	6694		Open cluster	15'	Scutum	9.5	5
M27	6853	Dumbbell nebula	Planetary nebula	8'×5.6'	Vulpecula	7.5	1.25
M28	6626		Globular cluster	11.2'	Sagittarius	8.5	18
M29	6913		Open cluster	7'	Cygnus	9	7.2
M30	7099		Globular cluster	3.5'	Capricorn	8.5	25
M31	224	Andromeda galaxy	Spiral galaxy	190'×60'	Andromeda	3.4	2,500
M32	221		Dwarf elliptical galaxy	8.7'×6.5'	Andromeda	10	2,900
M33	598	Triangulum galaxy	Spiral galaxy	70.8'×41.7'	Triangulum	5.7	2,810
M34	1039		Open cluster	35'	Perseus	6	1.4
M35	2168		Open cluster	28'	Gemini	5.5	2.8
M36	1960		Open cluster	12'	Auriga	6.5	4.1
M37	2099		Open cluster	24'	Auriga	6	4.6
M38	1912		Open cluster	21'	Auriga	7	4.2
M39	7092		Open cluster	32'	Cygnus	5.5	0.8
M40		Winnecke 4	Double star WNC4	0.8'	Ursa Major	9	0.5
M41	2287		Open cluster	38'	Canis Major	4.5	2.3
M42	1976	Orion nebula	H II region	65'×60'	Orion	4	1.6
M43	1982	De Mairan's nebula	H II region (part of the Orion Nebula)	20'×15'	Orion	7	1.6
M44	2632	Beehive cluster	Open cluster	95'	Cancer	3.7	0.6
M45		Pleiades	Open cluster	110'	Taurus	1.6	0.4
M46	2437		Open cluster	27'	Puppis	6.5	5.4
M47	2422		Open cluster	30'	Puppis	4.5	1.6

(continued)

The Messier list (continued)

Messier number	NGC number	Object name	Type of object	Dimensions	Constellation	Apparent magnitude	Distance (K LY)
M48	2548		Open cluster	54'	Hydra	5.5	1.5
M49	4472		Elliptical galaxy	10.2'×8.3'	Virgo	10	60,000
M50	2323		Open cluster	16'	Monoceros	7	3
M51	5194/5195	Whirlpool galaxy	Spiral galaxy	11.2'×6.9'	Canes Venatici	8.4	37,000
M52	7654		Open cluster	13'	Cassiopeia	8	7
M53	5024		Globular cluster	13'	Coma Berenices	8.5	56
M54	6715		Globular cluster	12'	Sagittarius	8.5	83
M55	6809		Globular cluster	19'	Sagittarius	7	17
M56	6779		Globular cluster	8.8'	Lyra	9.5	32
M57	6720	Ring nebula	Planetary nebula	230"×230"	Lyra	8.8	2.3
M58	4579		Barred spiral galaxy	5.9'×4.7'	Virgo	11	60,000
M59	4621		Elliptical galaxy	5.4'×3.7'	Virgo	11.5	60,000
M60	4649		Elliptical galaxy	7.4'×6'	Virgo	10.5	60,000
M61	4303		Spiral galaxy	6.5'×5.8'	Virgo	10.5	60,000
M62	6266		Globular cluster	15'	Ophiuchus	8	22
M63	5055	Sunflower galaxy	Spiral galaxy	12.6'×7.2'	Canes Venatici	8.5	37,000
M64	4826	Black Eye galaxy	Spiral galaxy	10.7'×5.1'	Coma Berenices	9	12,000
M65	3623		Barred Spiral galaxy	8.7'×2.5'	Leo	10.5	35,000
M66	3627		Barred spiral galaxy	9.1'×4.2'	Leo	10	35,000
M67	2682		Open cluster	30'	Cancer	7.5	2.25

Messier	NGC	Type	Common name	Size	Constellation	Magnitude	Distance
M68	4590	Globular cluster		11'	Hydra	9	32
M69	6637	Globular cluster		9.8'	Sagittarius	9	25
M70	6681	Globular cluster		8'	Sagittarius	9	28
M71	6838	Globular cluster		7.2'	Sagittarius	8.5	12
M72	6981	Globular cluster		6.6'	Aquarius	10	53
M73	6994	Asterism		2.8'	Aquarius	9	
M74	628	Spiral galaxy		10.5'×9.5'	Pisces	10.5	35,000
M75	6864	Globular cluster		6.8'	Sagittarius	9.5	58
M76	650/651	Planetary nebula	Little Dumbbell nebula	2.7'×1.8'	Perseus	10.1	3,4
M77	1068	Spiral galaxy		7.1'×6'	Cetus	10.5	60,000
M78	2068	Diffuse nebula		8'×6'	Orion	8	1.6
M79	1904	Globular cluster		8.7'	Lepus	8.5	40
M80	6093	Globular cluster		10'	Scorpius	8.5	27
M81	3031	Spiral galaxy	Bode's galaxy	26.9'×14.1'	Ursa Major	6.9	12,000
M82	3034	Starburst galaxy	Cigar galaxy	11.2'×4.3'	Ursa Major	9.5	11,000
M83	5236	Barred spiral galaxy	Southern Pinwheel galaxy	12.9'×11.5'	Hydra	8.5	10,000
M84	4374	Lenticular galaxy		6.5'×5.6'	Virgo	11	60,000
M85	4382	Lenticular galaxy		7.1'×5.5'	Coma Berenices	10.5	60,000
M86	4406	Lenticular galaxy		8.9'×5.8'	Virgo	11	60,000
M87	4486	Elliptical galaxy		8.3'×6.6'	Virgo	11	60,000
M88	4501	Spiral galaxy		6.9'×3.7'	Coma Berenices	11	60,000
M89	4552	Elliptical galaxy		5.1'×4.7'	Virgo	11.5	60,000
M90	4569	Spiral galaxy		9.5'×4.4'	Virgo	11	60,000
M91	4548	Barred Spiral galaxy		5.4'×4.3'	Coma Berenices	11	60,000
M92	6341	Globular cluster		14'	Hercules	7.5	26

(continued)

The Messier list (continued)

Messier number	NGC number	Object name	Type of object	Dimensions	Constellation	Apparent magnitude	Distance (K LY)
M93	2447		Open cluster	22'	Puppis	6.5	4.5
M94	4736		Spiral galaxy	11.2'×9.1'	Canes Venatici	9.5	14,500
M95	3351		Barred spiral galaxy	3.1'×2.9'	Leo	11	38,000
M96	3368		Spiral galaxy	7.6'×5.2'	Leo	10.5	38,000
M97	3587	Owl nebula	Planetary nebula	3.4'×3.3'	Ursa Major	9.9	2.6
M98	4192		Spiral galaxy	9.8'×2.8'	Coma Berenices	11	60,000
M99	4254		Spiral galaxy	5.4'×4.7'	Coma Berenices	10.5	60,000
M100	4321		Spiral galaxy	7.4'×6.3'	Coma Berenices	10.5	60,000
M101	5457	Pinwheel galaxy	Spiral galaxy	28.8'×26.9'	Ursa Major	7.9	27,000
M102	5866	Spindle galaxy	Spiral galaxy	4.7'×1.9'	Draco	10.7	50,000
M103	581		Open cluster	6'	Cassiopeia	7	8
M104	4594	Sombrero galaxy	Spiral galaxy	8.7'×3.5'	Virgo	9.5	50,000
M105	3379		Elliptical galaxy	5.4'×4.8'	Leo	11	38,000
M106	4258		Spiral galaxy	18.6'×7.2'	Canes Venatici	9.5	25,000
M107	6171		Globular cluster	13'	Ophiuchus	10	20
M108	3556		Spiral galaxy	8.7'×2.2'	Ursa Major	11	45,000
M109	3992		Barred spiral galaxy	7.6'×4.7'	Ursa Major	11	55,000
M110	205		Dwarf elliptical galaxy	21.9'×11'	Andromeda	10	2,200

The Caldwell catalog

Caldwell number	NGC number	Object name	Type of object	Dimensions	Constellation	Apparent magnitude	Distance (K LY)
C1	188		Open cluster	15'	Cepheus	8.1	4.8
C2	40		Planetary nebula	38"×35"	Cepheus	11.6	3.5
C3	4236		Barred spiral galaxy	21.9'×7.2'	Draco	9.7	7,000
C4	7023	Iris nebula	Open cluster and nebula	18'×18'	Cepheus	6.8	1.4
C5	IC342		Spiral galaxy	21.4'×20.9'	Camelopardalis	9.2	13,000
C6	6543	Cat's Eye nebula	Planetary nebula	20"	Draco	8.8	3
C7	2403		Spiral galaxy	21.9'×12.3'	Camelopardalis	8.9	14,000
C8	559		Open cluster	4.4'	Cassiopeia	9.5	3.7
C9	Sh2-155	Cave nebula	Nebula	50'×30'	Cepheus	7.7	2.8
C10	663		Open cluster	16'	Cassiopeia	7.1	7.2
C11	7635	Bubble nebula	Nebula	15'×8'	Cassiopeia	7	7.1
C12	6946		Spiral galaxy	11.5'×9.8'	Cepheus	9.7	18,000
C13	457	Owl cluster, E.T. cluster	Open cluster	13'	Cassiopeia	6.4	7.9
C14	869/884	Double cluster, H & X Persei	Open cluster	60'	Perseus	4.3	7.3
C15	6826	Blinking Planetary	Planetary nebula	27"×24"	Cygnus	9.8	2.2
C16	7243		Open cluster	21'	Lacerta	6.4	2.5
C17	147		Dwarf spheroidal galaxy	13.2'×7.8'	Cassiopeia	9.3	2,300
C18	185		Dwarf spheroidal galaxy	11.7'×10'	Cassiopeia	9.2	2,300
C19	IC5146	Cocoon nebula	Open cluster and nebula	12'	Cygnus	10	3.3

(continued)

The Caldwell catalog (continued)

Caldwell number	NGC number	Object name	Type of object	Dimensions	Constellation	Apparent magnitude	Distance (K LY)
C20	7000	North America nebula	Nebula	120'×100'	Cygnus	6	1.8
C21	4449		Irregular galaxy	6.2'×4.4'	Canes Venatici	9.4	10,000
C22	7662	Blue Snowball	Planetary nebula	37"	Andromeda	9.2	3.2
C23	891		Spiral galaxy	13.5'×2.5'	Andromeda	9.9	31,000
C24	1275	Perseus A	Seyfert galaxy	2.2'×1.7'	Perseus	11.6	230,000
C25	2419		Globular cluster	6'	Lynx	10.4	275
C26	4244		Spiral galaxy	16'×2.5'	Canes Venatici	10.6	10,000
C27	6888	Crescent nebula	Nebula	18'×12'	Cygnus	7.5	4.7
C28	752		Open cluster	75'	Andromeda	5.7	1.2
C29	5005		Spiral galaxy	5.8'×2.8'	Canes Venatici	9.8	6,000
C30	7331		Spiral galaxy	10.5'×3.7'	Pegasus	9.5	47,000
C31	IC405	Flaming Star nebula	Nebula	37'×10'	Auriga	6	1.6
C32	4631	Whale galaxy	Spiral galaxy	15.5'×2.7'	Canes Venatici	9.3	22,000
C33	6992	East Veil nebula	Supernova remnant	60'×8'	Cygnus	7	2.5
C34	6960	West Veil nebula	Supernova remnant	70'×6'	Cygnus	7	2.5
C35	4889		Elliptical galaxy	3'×2'	Coma Berenices	11.4	300,000
C36	4559		Spiral galaxy	10.7'×4.4'	Coma Berenices	9.8	32,000
C37	6885		Open cluster	7'×18'	Vulpecula	5.7	1.95
C38	4565	Needle galaxy	Spiral galaxy	15.9'×1.85'	Coma Berenices	9.6	32,000

C39	2392	Eskimo nebula, Clown Face nebula	Planetary nebula	48″×48″	Gemini	9.9	4
C40	3626		Spiral galaxy	2.7′×1.9′	Leo	10.9	86,000
C41	Mel25	Hyades	Open cluster	330′	Taurus	—	0.15
C42	7006		Globular cluster	2.8′	Delphinus	10.6	135
C43	7814		Spiral galaxy	5.5′×2.3′	Pegasus	10.5	49,000
C44	7479		Barred spiral galaxy	4.1′×3.1′	Pegasus	11	106,000
C45	5248		Spiral galaxy	6.2′×4.5′	Bootes	10.2	74,000
C46	2261	Hubble's Variable nebula	Nebula	2′	Monoceros	10	2.5
C47	6934		Globular cluster	8.4′	Delphinus	8.9	57
C48	2775		Spiral galaxy	4.3′×3.3′	Cancer	10.3	55,000
C49	2237	Rosette nebula	Nebula	80′×60′	Monoceros	9	4.9
C50	2244		Open cluster	24′	Monoceros	4.8	4.9
C51	IC1613		Irregular galaxy	16.2′×14.5′	Cetus	9	2,300
C52	4697		Elliptical galaxy	4.4′×2.8′	Virgo	9.3	76,000
C53	3115	Spindle galaxy	Elliptical galaxy	7.2′×2.5′	Sextans	9.1	22,000
C54	2506		Open cluster	7″	Monoceros	7.6	10
C55	7009	Saturn nebula	Planetary nebula	41″×35″	Aquarius	8.3	1.4
C56	246		Planetary nebula	3.8′	Cetus	8	1.6
C57	6822	Barnard's galaxy	Irregular galaxy	15.5′×13.5′	Sagittarius	9.3	2,300
C58	2360		Open cluster	13′	Canis Major	7.2	3.7
C59	3242	Ghost of Jupiter	Planetary nebula	25″	Hydra	8.6	1.4
C60	4038	Antennae galaxies	Spiral galaxy	5.2′×3.1′	Corvus	11.3	83,000
C61	4039	Antennae galaxies	Spiral galaxy	3.1′×1.6′	Corvus	13	83,000
C62	247		Spiral galaxy	21.4′×6.9′	Cetus	8.9	6,800

(continued)

The Caldwell catalog (continued)

Caldwell number	NGC number	Object name	Type of object	Dimensions	Constellation	Apparent magnitude	Distance (K LY)
C63	7293	Helix nebula	Planetary nebula	16'×28'	Aquarius	6.5	0.522
C64	2362		Open cluster and nebula	8'	Canis Major	4.1	5.1
C65	253	Sculptor galaxy, Silver Coin galaxy	Spiral galaxy	27.5'×6.8'	Sculptor	7.1	9,800
C66	5694		Globular cluster	3.6'	Hydra	10.2	113
C67	1097		Barred spiral galaxy	9.3'×6.3'	Fornax	9.2	47,000
C68	6729		Nebula	2.5'×2'	Corona Australis	9.7	0.424
C69	6302	Bug nebula, Butterfly nebula	Planetary nebula	3'	Scorpius	12.8	5.2
C70	300		Spiral galaxy	21.9'×15.5'	Sculptor	8.1	3,900
C71	2477		Open cluster	27'	Puppis	5.8	3.7
C72	55		Irregular galaxy	32.4'×5.6'	Sculptor	8.2	4,200
C73	1851		Globular cluster	11'	Columba	7.3	39.4
C74	3132	Eight Burst nebula	Planetary nebula	62"×43"	Vela	8.2	2
C75	6124		Open cluster	29'	Scorpius	5.8	1.5
C76	6231		Open cluster and nebula	15'	Scorpius	2.6	6
C77	5128	Centaurus A	Peculiar galaxy	25.7'×20'	Centaurus	7	16,000
C78	6541		Globular cluster	13'	Corona Australis	6.6	22.3
C79	3201		Globular cluster	18.2'	Vela	6.7	17
C80	5139	Omega Centauri	Globular cluster	36.3'	Centaurus	3.6	17.3

C	NGC/IC	Name	Type	Size	Constellation	Mag	Distance
C81	6352		Globular cluster	7'	Ara	8.1	18.6
C82	6193		Open cluster	15'	Ara	5.2	4.3
C83	4945		Spiral galaxy	20'×3.8'	Centaurus	9.5	17,000
C84	5286		Globular cluster	9'	Centaurus	7.6	36
C85	IC2391	Omicron Vela cluster	Open cluster	50'	Vela	2.5	0.5
C86	6397		Globular cluster	32'	Ara	5.6	7.5
C87	1261		Globular cluster	7'	Horologium	8.4	55.5
C88	5823	S Norma cluster	Open cluster	10'	Circinus	7.9	3.4
C89	6087		Open cluster	12'	Norma	5.4	3.3
C90	2867		Planetary nebula	12"	Carina	9.7	5.5
C91	3532		Open cluster	55'	Carina	3	1.6
C92	3372	Eta Carinae nebula	Nebula	120'×120'	Carina	6.2	7.5
C93	6752		Globular cluster	20.4'	Pavo	5.4	13
C94	4755	Jewel Box	Open cluster	10'	Crux	4.2	4.9
C95	6025		Open cluster	12'	Triangulum Australe	5.1	2.5
C96	2516		Open cluster	30'	Carina	3.8	1.3
C97	3766	Pearl cluster	Open cluster	12'	Centaurus	5.3	5.8
C98	4609		Open cluster	5'	Crux	6.9	4.2
C99		Coalsack nebula	Dark nebula	7°×5°	Crux		0.61
C100	IC2944	Lambda Centauri nebula, Running Chicken nebula	Open cluster and nebula	75'	Centaurus	4.5	6

(continued)

The Caldwell catalog (continued)

Caldwell number	NGC number	Object name	Type of object	Dimensions	Constellation	Apparent magnitude	Distance (K LY)
C101	6744		Spiral galaxy	20' × 12.9'	Pavo	9	34,000
C102	IC2602	Theta Carinae cluster, Southern Pleiades	Open cluster	50'	Carina	1.9	0.492
C103	2070	Tarantula nebula	Open cluster and nebula	40' × 25'	Dorado	8.2	170
C104	362		Globular cluster	12.9'	Tucana	6.6	27.7
C105	4833		Globular cluster	13.5'	Musca	7.3	19.6
C106	104	47 Tucanae	Globular cluster	30.9'	Tucana	4	14.7
C107	6101		Globular cluster	11'	Apus	9.3	49.9
C108	4372		Globular cluster	19'	Musca	7.8	18.9
C109	3195		Planetary nebula	40" × 35"	Chamaeleon	11.6	5.4

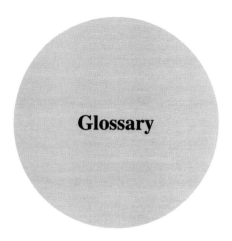

Glossary

Acclimation Allowing a telescope to take on the ambient air temperature in order to reduce the need for constant refocusing.

Alignment See **Polar alignment**.

Alt-az Abbreviation for the altitude-azimuth method of aligning a telescope so that it steadily tracks objects in the night sky compensating for the rotation of Earth.

Altitude A coordinate that determines how high in the sky a telescope needs to point in order to locate an object in space.

Aperture The diameter of a telescope's light-gathering front lens.

Apochromatic A short refracting telescope specifically designed to be used for wide-field CCD imaging.

Arc-degrees; Arc-minutes; Arc-seconds A unit of measurement of the celestial sky where the sphere of sky around the planet is divided into 360 arc-degrees. Arc-degrees are further divided into 60 arc-minutes, which are in turn divided into 60 arc-seconds.

Astrometry The determination of the coordinates of celestial objects using the known coordinates of reference stars and the measured distance of the object from these stars.

Auto-guiding Using a second telescope and CCD imager to guide the movements of the primary imaging telescope. The guide 'scope helps keep the imaging telescope precisely pointed on the imaging subject.

Azimuth A coordinate that determines what compass direction a telescope needs to point to in order to locate an object in space.

Barlow lens An eyepiece attachment used to increase the magnification of objects through a telescope.

L.A. Kennedy, *One-Shot Color Astronomical Imaging*, Patrick Moore's
Practical Astronomy Series, DOI 10.1007/978-1-4614-3247-0,
© Springer Science+Business Media New York 2012

Bayer array An alternating pattern of colored pixel filters on an imaging chip used to create a multiple-color image from a single exposure.

Bias frames A very short covered exposure used to correct images by reducing noise inherent in the CCD imager itself. Bias frames are used only when combined with dark frames that have different exposure lengths than the images being corrected.

Black point The spot assigned to a histogram slider that defines which pixel values will be displayed as completely black in an image. The pixel values between the black and white points are then allocated to the limited number of color variations that can be displayed on a digital screen.

Blurring An image-processing technique used to smooth out noisy image backgrounds by blending pixels together.

Bubble level A small circular device used to ensure your telescope's base is set up perfectly flat and even, which enhances telescope tracking and imaging quality.

Caldwell list A list of 109 interesting night-sky objects that are excellent subjects for imaging. This list was developed by Sir Patrick Caldwell-Moore as a supplement to the list of Messier objects.

Calibration Using baseline exposures of inherent noise and random flaws to remove these unwanted features from your final images.

CCD A charge-coupled device, also known as an imaging camera or imager. The name stems from the way in which rows of pixels are manufactured, or coupled together, in order to pass their electrical charge down the row for readout by the imaging control program.

Centroid The calculated center of the image of a star being used by the imaging control program for auto-guiding, measurement, or other purposes.

Clip A mathematical or statistical method used to eliminate pixels whose measurements fall outside of the assigned range of acceptable values.

Color balancing A program feature used to adjust the input values assigned to each color being used to make up a composite color image.

Color range tool A program feature in Photoshop™ used to make a selection of all pixels in an image whose measurements fall within the assigned range of values.

Combine A method of integrating data values from multiple images into one composite image with an increased signal-to-noise ratio; also known as "stacking".

Correction gain A setting in an auto-guiding control program that defines what percentage of a needed correction move should be applied at one time. Limiting the correction move to a fraction of the complete move reduces the chance of over-correction and eliminates the need for a counter-correction in the opposite direction.

Critical focus zone A very small range in the path of the light coming through a telescope lens where all of the light converges into one area.

Curves A program feature in Photoshop™ used to make multiple non-linear histogram adjustments for an image. The flexibility of the Curves feature allows many small adjustments to be made at different points along the histogram curve, thereby providing more control when enhancing particular areas and features of an imaging target.

Dark frame A covered exposure used to correct unwanted features in your final images. Dark frames are used to eliminate predictable noise created by heat stemming from a CCD camera's internal circuitry.

Dark noise Unwanted noise in images that show up as very bright or "hot" pixels. This noise is generated by the heat coming from a CCD camera's internal circuitry even when the imaging chip is not being exposed to a light source.

Declination An imaginary plane that extends from Earth's equator outward, infinitely, into space.

Deconvolution An image-processing technique used to enhance images that are slightly out of focus or not quite as clear and crisp as they otherwise could be.

Dew protector A small electric heating pad that wraps around the lenses on a telescope, thereby heating them just enough to keep dew from forming on the glass surfaces.

Dew shield An extension that fits onto the light-gathering end of a telescope used to keep the colder air off the surface of the lens longer, delaying dew formation.

Digital development An image-processing technique used for image enhancement that incorporates both a sharpening routine and a non-linear histogram stretch. This is accomplished using a processing filter known as a kernel, which is a type of mathematical matrix that tells the processing program exactly how to process the pixels surrounding an area of interest in your image.

Drift alignment A procedure used to ensure that a telescope's axes are correctly aligned to the celestial pole in order to better track objects while compensating for the rotation of Earth.

Dust mote A flaw in an image caused by an extremely magnified piece of dirt or dust that is lying on one of the imaging lenses or other glass surfaces in the imaging path.

Dynamic range The difference between the brightest and the faintest areas of a single image that can be captured on a given CCD camera's imaging chip. The higher the dynamic range of an imaging chip, the greater the differences that can be captured in a single image.

FFT An abbreviation for Fast Fourier Transform, a low-pass type of filter used in the Digital Development processing routine that tends to give an image more sharpening with less non-linear histogram stretching than other filter types.

Field of view (FOV) The area of the sky that can be seen through an eyepiece or can be imaged by a CCD camera's imaging chip.

Filter wheel A mechanical device that houses multiple colored or emission filters. The device allows the user to switch filters by rotating the wheel rather than having to tear apart the imaging train in order to install a new filter for imaging.

Filters A glass lens used in the imaging train manufactured with a special coating that allows only certain wavelengths of light to pass through the lens.

FITS An abbreviation for Flexible Image Transport System, a type of file format specifically designed for use in saving astronomical and other scientific images.

Flat-field frame An image taken of an evenly illuminated surface that is then used to correct or eliminate dust motes and other flaws caused by dirt or obstructions on the imaging surfaces; also known as a light frame.

Flexure The uncoordinated movement of one telescope relative to another when using both an imaging and a guiding telescope simultaneously.

Focal length The distance the light path travels from the light-gathering lens to the focal point of the eyepiece or the imaging chip.

Focal ratio The focal length of the telescope divided by the aperture of the primary lens. Typically categorized as fast (smaller ratio) and slow (larger ratio), a telescope with a faster focal ratio is generally easier to use for astro-imaging than one with a slower focal ratio. However, a slower focal ratio offers more resolution and a higher level of magnification.

Focal reducer A lens attachment designed to decrease the magnification of an imaging setup, thereby increasing the field of view that can be imaged.

Focus The art of getting the imaging chip into the precise spot where the light rays converge into a single path. This provides the clearest, sharpest images.

Framing The art of getting your imaging subject to appear on your image in exactly the precise spot to best display the target aesthetically.

Go-To The capability of today's computerized telescopes to accurately point themselves at specific locations in the sky where objects of interest are located.

Grayscale A type of "black and white" image that actually consists of a wide range of differing values of gray pixels.

Guide star A relatively bright star located in the field of view of a guide 'scope or off-axis guider that is used to keep the telescope pointed at precisely the same spot in the sky.

Hartmann mask A focusing aid consisting of a flat piece of cardboard or other material with several holes cut into it. The mask is placed on the light-gathering end of a telescope as an aid in obtaining rough focus.

Histogram A visual graph showing the amount of data (photons) present in each pixel across the image as a whole; the graph indicates the number of pixels with the same photon count from zero through the maximum (full-well) count of the CCD camera's imaging chip.

Histogram stretch Adjustment of the amount of the histogram being displayed so that only the desired pixel values are allocated to the available display values on a digital screen.

Hot pixels Flaws in an image where pixels are extremely bright relative to the actual values that should be present. Caused by a phenomenon known as dark noise, hot pixels can usually be corrected through the application of dark frames during the calibration process.

Image reduction The process of correcting flaws in an image, or set of images, caused by dirt or dust on the surface lenses or the inherent noise generated by electronic circuitry in the CCD camera itself.

Image scale The measurement of the area of sky a single pixel on an imaging chip will capture given the focal length of the telescope it is being used with.

Imaging chip The light-sensitive computer chip at the heart of a CCD camera that gathers photons as light reaches the surface of the chip.

Imaging train The line of equipment pieced together to be used in capturing images with a CCD camera.

Interpolation The process by which a computer program populates the values of a single color image using the multi-colored pixel data in a Bayer array.

Iterations A series of image-processing routines run multiple times.

Layers tool A program feature in *Photoshop™* used to combine multiple images into a single composite image. This tool is often used to highlight certain features of an imaging subject in each layer so that all of these features are highlighted in the composite image.

Light frame See **Flat-field frame**.

Light pollution Excessive lighting from unshielded streetlamps, neon lights, and other light sources that fill the sky with unwanted photons. This stray light blocks our view of many stars in the night sky and can cause a CCD imaging chip to become saturated with light; also known as skyglow.

Lightbox A portable device used to take flat-field images. The box is constructed with an internal light source, which evenly illuminates a white Plexiglas sheet. The box slips over the light-gathering end of a telescope, and images are taken of the evenly illuminated surface to capture a light frame.

Lighted reticule eyepiece A specialized eyepiece that contains an internal light source used to illuminate a set of crosshairs on the lens. Primarily used for manual guiding with an off-axis guider or for performing the drift alignment procedure.

LPS Abbreviation for Light Pollution Suppression. A lens filter used in the imaging train that has been manufactured with a special coating that blocks wavelengths in the light spectrum of most man-made sources of light, but allows natural light from stars to pass through to the imaging chip.

LRGB image A composite image made by combining individual red, green, and blue filtered images along with an unfiltered full-spectrum image, known as a luminance image.

Luminance image An unfiltered full-spectrum "black and white" or grayscale image used to help bring out fine details in a composite LRGB image.

Messier list A list of 110 interesting deep-sky objects developed by French comet-hunter Charles Messier in the late 1700s. The list was originally created to help him and other comet hunters avoid confusing these objects with comets during sky searches.

Micron A unit of measurement equal to one-millionth of a meter.

Mosaic A combination of multiple images lying side by side, with slightly overlapping edges, so that they form one single larger image.

Narrow-band filter A glass lens with a special coating that only allows certain wavelengths of light to pass through. The light allowed to pass through corresponds to the specific wavelengths given off by certain atoms such as hydrogen-alpha (Hα), triply-ionized oxygen (OIII), sulfur (SII), and hydrogen-beta (Hβ).

Noise Unwanted "false" data or uncertainty in the brightness levels of the pixels in an image.

Non-linear histogram stretch Adjustment of the shape of the histogram so that a particular range of pixel values are allocated to a disproportionate amount of the available display values on a digital screen.

Normalization Adjustment made to multiple images before combination so that the background pixel values are roughly equal, thereby adjusting all pixel values to the correct relative range.

Off-axis guider A device attached in the imaging train that houses a lighted reticule eyepiece at a 90° angle relative to the light path. An internal mirror is used to pick off a portion of the light path and divert it to the eyepiece for use in manual guiding.

PEC Abbreviation for Periodic Error Correction. A feature available on some telescopes that automatically compensates for flaws in the drive gear that can cause inaccuracies in tracking objects.

Photometry The measurement of magnitude, or how much light is coming from a specific star or other object.

Photons Particles of light that make up a light wave. These particles are collected by a CCD camera's pixels. The pixel is assigned a display value based on how many particles are collected during the length of the exposure.

Pixel A shortened term for "picture element." The light-sensitive collectors laid out in rows and columns that make up a CCD imaging chip or the small dots that make up the display screen on a digital device.

Pixel size The physical measurement of the size of the light-sensitive collectors that make up a CCD imaging chip. Pixel size is used to determine various calculations such as field of view and image scale.

Polar alignment A procedure used to make sure the central axis of a telescope is perfectly in line with Earth's axis so that it can more easily track objects across the sky as Earth rotates.

PSF Abbreviation for Point Spread Function. The mathematical definition of how the light from a point source, such as a star, gets spread out across the pixels on an imaging chip.

Quick mask A feature in *Photoshop*™ that allows you to quickly see which areas of an image are affected when making selections with one of the various selection tools available.

Reflector A type of telescope that uses mirrors to increase the focal length of the light path well beyond the length of the telescope tube.

Refractor A type of telescope in which the light path runs straight through the telescope's tube from the light-gathering lens to the eyepiece.

RGB image A composite image made by combining individual red, green, and blue filtered images.

Right ascension An imaginary 360° sphere of coordinates that extends infinitely into space. These coordinates, measured in "hours" and "minutes" are used to locate the position of planets, stars, and other objects in the universe from any point on Earth.

SCT Abbreviation for Schmidt-Cassegrain Telescope, a type of telescope that is ideally suited to do astro-imaging with a CCD camera.

Scintillation The twinkling of stars or other objects due to atmospheric turbulence.

Sharpening An image-processing technique used to enhance images by increasing the differences in nearby pixels, thereby enhancing the contrast between the different features in an image.

Sigma clip A statistical method used to eliminate pixels whose measurements fall outside the assigned range of acceptable values.

Signal The desired parts of an image containing reasonably certain, relatively accurate pixel data.

Signal-to-noise ratio The ratio of desired, relatively accurate pixels compared to the undesirable, statistically inaccurate pixels in an image. Defined as the number of "good" pixels in an image divided by the number of "bad" pixels in the same image; any effort that can be made to increase this ratio (by decreasing noise or increasing signal, or both) will result in a much better final image.

Skyglow See **Light pollution**.

Smoothing See **Blurring**.

Stacking Mathematically combining multiple images of the same object into a composite image. This process increases signal and reduces noise, thereby increasing the signal-to-noise ratio, which results in more statistically accurate pixel values.

Thermo-electronic cooling A cooling feature of modern CCD cameras that uses circuitry and manufacturing design to reduce the working temperature of the camera, thereby reducing the amount of dark noise in an image.

Unsharp masking An image processing technique used to enhance images by determining which pixels are blurry (unsharp) and then correcting (masking) these blurry pixel values in the processed image.

USB An abbreviation for Universal Serial Bus. A type of cable used to connect a CCD camera to a computer. The cable has a specific type of connector that couples with a specific type of outlet (USB port) available on most computers.

Vibration control pads A set of gel-filled pads that sit under the legs of a telescope's tripod in order to dampen external vibrations that can cause unwanted movement of the telescope tube during imaging.

Wedge A triangular-shaped base designed for use with a Schmidt-Cassegrain Telescope to enable the telescope's axis to be lined up with the axis of Earth's rotation.

White point The spot assigned to a histogram slider that defines which pixel values will be displayed as completely white in an image. The pixel values between the black and white points are then allocated to the limited number of color variations that can be displayed on a digital screen.

Wind screen A portable set of panels that can be set up around a telescope to keep wind from causing unwanted movement of the telescope tube during imaging.

Zenith The point in the sky directly overhead. Typically the clearest spot in which to image since the light from objects overhead travels the shortest path through Earth's turbulent atmosphere.

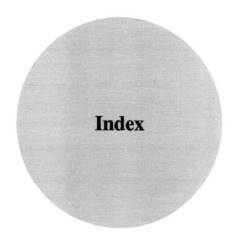

Index

A
Acclimation, 34, 181
Altitude, 44, 64–68, 181
Asteroid, 165–166
Astrometry, 164, 165, 181
Atmospheric extinction, 165
Auto-guiding, 20, 38–40, 62, 64, 89, 99–111,
    161, 181
Average combine, 95, 122, 123
Azimuth, 64–69, 181

B
Backlash, 111
Bayer array, 8, 9, 104, 129, 160, 181
Blinking, 47, 48, 165, 175
Blurring, 148–151, 153, 182
Brightness, 48, 78, 93, 99, 102, 114,
    135, 138, 142, 144, 146, 151,
    161, 164, 165

C
Calibration, 31, 32, 34, 35, 40, 41, 85, 87, 89,
    92, 93, 95, 96, 120, 128–130, 182
Clipping, 143
Color balance, 131, 132, 135–137, 148
Colored filters, 7–9, 22, 120, 128, 160

Color-range, 135, 146, 148, 150, 151, 182
Color range selection tool, 135, 146,
    148, 150, 151
Comet, 44, 165–166
Correction gain, 110, 182
Critical focus zone, 78–81, 153, 182
Curves, 56, 57, 116, 119, 138, 141,
    143, 145, 146, 159, 164, 165,
    167, 182

D
Dark current, 35, 86–88, 95, 96, 122
Dark frames, 34–36, 40, 41, 61, 63, 82,
    85–89, 95–97, 99, 119, 120, 122,
    128, 129, 183
Declination (Dec.), 66–70, 107, 165, 183
Deconvolution, 153–156, 183
Dew protector, 28–30, 183
Dew shield, 28, 29, 63, 163
Diffraction mask, 76, 77
Diffraction spikes, 76, 77
Digital development, 151–155, 183
Digitalspaceimages.com, 27, 32, 72,
    95, 168
Drift alignment, 27, 35, 66–70, 183
Dynamic range, 95, 107, 113, 118, 125, 133,
    134, 140, 160, 183

L.A. Kennedy, *One-Shot Color Astronomical Imaging*, Patrick Moore's
Practical Astronomy Series, DOI 10.1007/978-1-4614-3247-0,
© Springer Science+Business Media New York 2012

E
Exposure times, 78, 95, 99–102
Eyedropper tool, 150

F
Feathering, 146, 150
Field of view (FOV), 20, 24, 39, 50–57, 61,
    63, 65, 67, 68, 71–73, 76, 82, 93, 101,
    106, 107, 109, 183
Filters, 7–9, 22, 30, 36, 50, 61–63, 93, 100,
    105, 120, 126, 128, 131, 133, 146, 152,
    153, 159–161, 183
Filter wheel, 160, 183
FITS. See Flexible Image Transport System
    (FITS)
Flat field, 31, 92–95, 97, 122, 129
Flat-field frames, 92–95, 97, 129
Flexible Image Transport System (FITS), 103,
    104, 137, 183
Focal length, 18, 19, 24, 36, 50–55, 107, 109,
    111, 184
Focal ratio, 19, 20, 52–55, 78, 107, 111, 184
Focus, 1, 4, 5, 8, 10, 11, 18, 27, 30, 31, 34,
    37–41, 51, 57, 62, 63, 67, 71–83, 92,
    99, 108, 146, 153, 160, 184
Focusing, 11, 30, 31, 34, 37, 38, 57, 63,
    71–83, 160
Focus lock, 72, 73, 78
FOV. See Field of view (FOV)

G
Gamma adjustments/setting, 135
Go-to, 13, 18, 36, 44, 47, 64, 82, 184
Guide scope, 106–109
Guide star, 38, 39, 66, 67, 105–111, 161, 184
Guiding, 20, 38–40, 62, 64

H
Hartmann mask, 30, 31, 37, 38, 72–74, 184
Histogram, 82, 95, 101, 113–126, 131, 134,
    137–146, 151–153, 161, 163, 184
Histogram stretch, 131, 138–146,
    151–153, 184
Hot pixels, 36, 85, 87–89, 165, 184
Hue, 136

I
Image reduction, 85, 89, 95, 96, 102, 105, 184
Image scale, 24, 36, 50–57, 139, 140, 163,
    165, 166, 184

Imaging train, 7, 8, 20, 30, 31, 36, 40, 50,
    53–55, 57, 59, 61–63, 71, 82, 85, 87,
    159, 160, 184
Individual color components, 93, 95, 104, 105,
    120, 126, 128–137

J
Journal, 56, 57, 73

L
Layers, 136–137, 146, 151, 163, 185
Levels, 14, 15, 18, 20, 22, 25, 28, 34, 55,
    60, 61, 66–68, 92, 93, 96, 102, 118,
    119, 125, 129–131, 133, 137–139,
    141–143, 146, 148, 152, 153,
    157–159, 168
Light box, 31–32, 94, 95, 185
Lighted-reticule eyepiece, 27, 35, 39, 64, 67,
    69, 105, 185
Light frames, 31, 40–41, 57, 85, 89–95, 119,
    120, 160, 195
Light pollution, 5, 14, 20, 30, 36, 39, 61–63,
    99, 100, 103, 114, 119, 133, 138, 139,
    165, 185
Light pollution suppression (LPS), 30, 61, 82,
    100, 185
Linear histogram stretch, 139, 140
Logbook, 56, 57
LPS. See Light pollution suppression (LPS)
LRGB, 93, 126, 136, 137, 148, 151, 185
Luminance, 9, 10, 93, 126, 132, 136–137,
    151, 185

M
Masking, 146, 148, 187
Median combine, 96, 104, 122, 123
Meteor, 101, 104, 165
Minor planet, 166, 169
Mirror shift, 72, 81, 82, 95
Mosaic, 161–163, 185
Motorized focuser, 73, 76, 79, 81
Mounts, wedge, 21, 61, 66, 69
Multiple exposures, 8, 41, 119, 126, 154

N
Narrow-band filter, 159–160, 185
Noise, 6, 9, 35, 39, 86, 87, 95, 96, 100,
    102–103, 105, 119–121, 123, 125, 126,
    128, 129, 133, 137, 146–151, 153, 154,
    158, 165, 185

Non-linear histogram stretch, 131, 138,
    141–146, 151–153, 185
Normalization, 186
Nova search, 164–165

O
Off-axis guider, 39, 105, 106, 186

P
PEC. *See* Period error correction (PEC)
Period error correction (PEC), 111, 186
Photometry, 164, 186
Photons, 1–7, 86, 87, 99, 101, 113, 116, 118,
    132, 159, 186
Photoshop, 135, 136, 143, 145, 146, 148, 150,
    151, 161, 163
Pixels, 4, 23, 35, 50, 65, 71, 85, 100, 113,
    127, 165, 186
Point spread function (PSF), 153, 155
Polar alignment, 21, 27–28, 34, 35, 57, 59–61,
    64–70, 99, 102, 105, 110, 111, 159, 186
Polygonal Lasso tool, 150
PSF. *See* Point spread function (PSF)

Q
Quick mask, 150, 186

R
RGB, 93, 126, 133, 136–137, 148, 160, 186
Right ascension (R.A.), 66, 67, 69, 70, 107,
    165, 186

S
Saturation, 5, 100, 102, 103, 119, 135, 136
Saving images, 37, 103–105,
    120, 137
Scintillation, 27, 186
Selection inversion, 146
Sharpening, 127, 136, 146–149,
    151–154, 186
Sigma clip, 123–125, 187
Signal-to-noise ratio (S/N), 102
Skyglow, 14, 61, 114, 116, 187
Smoothing, 148–151, 153, 187
Solar imaging, 160–161
Stacking, 6, 89, 100, 113–126, 128–131, 137,
    154, 157, 159, 187
Supernova search, 90, 164–165

T
Telescopes types of, 10–12
Thermal noise, 86, 87, 96
Time lapse, 164

U
Unsharp mask, 146, 148, 149, 187

V
Vibration, 14, 25–27, 60, 61, 73, 120
Vibration control, 25–27, 60, 61, 187

W
Wind screen, 187

16665076R00115

Made in the USA
Lexington, KY
04 August 2012